University of California Publications

ZOOLOGY
Volume 122

Vertebral Morphology, Alternation of Neural Spine Height, and Structure in Permo-Carboniferous Tetrapods, and a Reappraisal of Primitive Modes of Terrestrial Locomotion

Stuart Shigeo Sumida

University of California Press

Vertebral Morphology, Alternation of Neural Spine Height, and Structure in Permo-Carboniferous Tetrapods, and a Reappraisal of Primitive Modes of Terrestrial Locomotion

Stuart Shigeo Sumida

UNIVERSITY OF CALIFORNIA PRESS
Berkeley • Los Angeles • Oxford

UNIVERSITY OF CALIFORNIA PUBLICATIONS IN ZOOLOGY

Editorial Board: Peter B. Moyle, James L. Patton,
Donald C. Potts, David S. Woodruff

Volume 122
Issue Date: August 1990

UNIVERSITY OF CALIFORNIA PRESS
BERKELEY AND LOS ANGELES, CALIFORNIA

UNIVERSITY OF CALIFORNIA PRESS, LTD.
OXFORD, ENGLAND

© 1990 BY THE REGENTS OF THE UNIVERSITY OF CALIFORNIA
PRINTED IN THE UNITED STATES OF AMERICA

Library of Congress Cataloging-in-Publication Data

Sumida, Stuart Shigeo.
    Vertebral morphology, alternation of neural spine height, and structure in Permo-Carboniferous tetrapods, and a reappraisal of primitive modes of terrestrial locomotion / Stuart Shigeo Sumida.
    p.   cm. — (University of California publications in zoology; v. 122)
    Includes bibliographical references (p.)
    ISBN 0-520-09755-6 (alk. paper)
    1. Vertebrates, Fossil. 2. Vertebrates—Locomotion. 3. Spine. 4. Paleontology—Permian. 5. Paleontology—Carboniferous. 6. Animal locomotion. 7. Anatomy, Comparative. I. Title. II. Series.
QE841.S86 1990
567—dc20   90-10885
                                                                CIP

The paper used in this publication meets the minimum requirements of American National Standard for Information Sciences—Permanence of Paper for Printed Library Materials, ANSI Z39.48-1984. ∞

To those people who taught me not to take myself seriously,
but to take what I do seriously: Donna Michiko Oye, Bill M. Ryusaki, and most
especially my parents, Shigeo Sumida, M.D. and Therese Sumida, M.A.

Contents

List of Illustrations, vii
Acknowledgments, ix
Abbreviations, xi
Abstract, xiv

1. INTRODUCTION   1
      Generalized Vertebral Structure in Tetrapods, 3
      Methods, 4
         Structural Context, 4
         Phylogenetic Context, 5

2. THE FAMILY CAPTORHINIDAE   6
      Captorhinus aguti, 8
      Eocaptorhinus laticeps, 19
      Labidosaurus, 21
      Protocaptorhinus, 29
      Captorhinikos, 33
         Captorhinikos chozaensis, 34
         Captorhinikos parvus, 35
         Captorhinikos valensis, 38
         Captorhinikos: Summary, 38
      Labidosaurikos, 38
      The Family Captorhinidae: Summary, 40

3. SEYMOURIAMORPHA AND DIADECTOMORPHA   42
      Seymouriamorpha, 43
         Seymouriidae: *Seymouria*, 43
      Diadectomorpha, 49
         Limnoscelidae, 49
            Limnoscelis, 49
         Tseajaiidae: *Tseajaia*, 52
         Diadectidae, 54
      Seymouriamorpha and Diadectomorpha: Summary, 58

4. THE MICROSAURIA AND OTHER "LEPOSPONDYLS"   59
Morphological Characteristics of the Microsauria, 59
Pantylidae: *Pantylus cordatus*, 60
Ostodolepidae, 63
    Ostodolepis brevispinatus, 63
    Pelodosotis elongatum, 64
    Micraroter, 65
Trihecatontidae: *Trihecaton howardinus*, 67
Molgophidae, 69
    Pleuroptyx clavatus, 70
    Molgophis macrurus, 71
Microsauria: Summary, 72

5. THE PELYCOSAURIA: *VARANOSAURUS*   73
Varanosaurus: Summary, 79

6. ARAEOSCELIDIA   81
Araeoscelis, 81
Zarcasaurus tanyderus, 86
Petrolacosaurus kansensis, 89
Araeoscelidia: Summary, 92

7. DISCUSSION   93
Functional Analysis, 93
    Articular Surfaces of the Vertebrae, 93
    Muscular Contributions to Movement, 95
    Other Mechanical Considerations, 99
Variability in Patterns of Alternation, 101
Phylogenetic Considerations, 102
    Alternation as a Character, 102
        Phylogenetic Applications, 103
        Phylogenetic Utility at the Familial Level, 104
Summary, 105
Conclusions, 105

Appendix: Specimens Examined, 107
Literature Cited, 115
Plates, 131

List of Illustrations

Figures

1. Vertebral structure in Permo-Carboniferous tetrapods, 2
2. Hypothesis of relationships of taxa in study, 3
3. *Captorhinus aguti*, dorsal view, 7
4. Atlas-axis complex of *Captorhinus aguti*, 10
5. Occiput of *Captorhinus aguti*, 11
6. Epaxial musculature of *Captorhinus aguti*, 12
7. Dorsal vertebrae of *Captorhinus aguti*, 13
8. Dorsal vertebrae of *Captorhinus aguti*, anterior view, 15
9. Axial musculature of *Captorhinus aguti*, 16
10. Sacral and caudal vertebrae of *Captorhinus aguti*, 18
11. Vertebral columns, *Eocaptorhinus* and *Captorhinus*, 20
12. *Labidosaurus hamatus*, dorsal view, 22
13. *Labidosaurus hamatus*, predominantly dorsal view, 24
14. Vertebrae of *Labidosaurus hamatus*, 27
15. Dorsal vertebrae of *Protocaptorhinus pricei*, 31
16. Caudal and costal elements of *Protocaptorhinus pricei*, 33
17. Dorsal vertebrae of *Captorhinikos chozaensis*, 34
18. Dorsal vertebrae of *Captorhinikos parvus*, 36
19. *Labidosaurikos* vertebrae, 39
20. Anterior dorsal vertebrae of *Seymouria*, 46
21. Mid-dorsal vertebrae of *Seymouria*, 48
22. Dorsal vertebrae of *Tseajaia campi*, 53
23. Vertebrae of diadectids, 57
24. Microsaurian vertebrae, 61
25. Axial structures of molgophid amphibians, 71
26. Presacral vertebrae of *Varanosaurus acutirostris*, 75
27. Sacral and caudal elements of *Varanosaurus acutirostris*, 78
28. *Araeoscelis gracilis* and *Petrolacosaurus kansensis*, 85
29. Vertebrae of *Zarcasaurus tanyderus*, 88
30. Articular surfaces of tetrapod vertebrae, 94

31. Determination of the axis of vertebral rotation, 95
32. Movement of vertebrae, 98
33. Factors affecting stability of a primitive reptile, 100

Plates

1. *Captorhinus aguti*, photomicrographs of thin sections, 133
2. Vertebral structures in *Labidosaurus hamatus*, 135
3. Axial structures in *Limnoscelis*, 137

Acknowledgments

First and foremost, I extend special thanks to Peter P. Vaughn, my dissertation advisor, for his patient advice (scientific, academic, and beyond) regarding this study. Thanks are also due Drs. Marshall Urist, Louis Goldberg, Thomas Howell, and especially Don Buth, whose advice and guidance are truly appreciated.

The support and encouragement of many people have helped this work to its completion. Most notably, Dr. Everett C. Olson has shared with me his vast knowledge of all aspects of paleontology, and has provided enthusiastic counsel on all aspects of my work. Further, Dr. Olson's pioneering work on the evolution of the tetrapod vertebral column provided an important inspiration and source of insight to the work presented here. Dr. David S Berman reviewed the entire manuscript, willingly gave me access to his unpublished data, discussed my work at length, and extended his hospitality to me during my visit to the Carnegie Museum of Natural History. Drs. Donald Brinkman, David Eberth, Robert Holmes, and Nicholas Hotton III freely shared with me their ideas, both published and unpublished. Drs. David S Berman, John Bolt, Eugene Gaffney, Nicholas Hotton III, Farish Jenkins, John Ostrom, and Hans-Peter Schultze allowed me to examine materials under their care.

Financial support for this work was provided, in part, by grants from the K. P. Schmidt Fund at the Field Museum of Natural History, short-term visitor's grants from the American Museum of Natural History and National Museum of Natural History, a UCLA Chancellor's dissertation fellowship, and a UCLA Department of Biology dissertation fellowship.

Al Myrick generously allowed me to use his petrographic microscope for the photomicrographs, and Dr. Diane Riska provided continual advice on all matters photographic. Special thanks are due Dr. Takeo Susuki for giving me access to his darkroom and equipment, and his hours of patient instruction on their use. The number of figures in this study attests to his interest and generosity. Drs. Ove Hoegh-Goldburgh, Randall D. Orton, and Walt Rainboth of UCLA provided critical support in the preparation of the final-copy version of the manuscript. Richard Sue deserves special recognition for aid in the monumental job of organizing the original manuscript.

Charles Solomon lent his vast writing experience, good humor, and thorough yet patient skills as an editor. Through his efforts, the contents of this manuscript finally began to resemble the English language. These acknowledgments are the only pages he has not read before publication.

Space does not permit me to list the actions of many who supported and helped this project to its completion. I trust that those listed will understand the acknowledgements of their contributions: Mark Abe, Sophie Bunce, Ben Crabtree, Anthony Dohi, Tammy Joe, Eric Lew, Eric Lombard, Karen Martin, Ron Matson, Mary Murakawa-Lew, Robert Nagy, David Nakamura, Ethan Nasreddin-Longo, Gregory Shimizu, Claire Sumida, Krista Sumida, Whitney Sumida-Davis, Doris Sue, Gordon Sue, Lawrence Sue, Lola Sue, Kathy Sue-Yamane, Alexandra and David Swafford, Sharon Takase, Susan Takei, David Toyofuku, Judith Verbeke, Myles Wakiyama, Tom White, Jeanie Woo, Jocelyn Yamadera, Michael Yamane, and Daven Yokota. Special thanks must be accorded to Dr. Fasha Liley.

Finally, for those to whom this dissertation is dedicated: to my wife Donna Michiko Oye, who has known me through the tenure of this project and has patiently endured my preoccupation with it; to Bill Ryusaki, who after my parents was the most important instructor that I ever had; and to my parents, Shigeo Sumida and Therese Sumida, who supported and encouraged my preoccupation with dirt, rocks, and bones. To these people, my deepest gratitude.

Abbreviations

Abbreviations Used in Text

The following institutional acronyms preceding catalogue numbers are used in the text:

AMNH	American Museum of Natural History; New York, New York
BSPHM	Bayerische Staatssammlung fur Palaontologie und Historische Geologie, Munich, Germany
CM	Carnegie Museum of Natural History, Pittsburgh, Pennsylvania
FMNH	Field Museum of Natural History, Chicago, Illinois
KUVP	University of Kansas Museum of Natural History; Lawrence, Kansas
NTMVP	Navajo Tribal Museum, Window Rock, Arizona; materials on extended loan to the Carnegie Museum of Natural History; Pittsburgh, Pennsylvania
MCZ	Museum of Comparative Zoology, Harvard University, Cambridge, Massachusettes
OUSM	Stovall Museum of Science and History, University of Oklahoma, Norman
UCLA VP	University of California, Los Angeles, Vertebrate Paleontology Collections
UCMP	University of California Museum of Paleontology, Berkeley
USNM	United States National Museum of Natural History, Smithsonian Institution, Washington, D.C.
UT	University of Texas, Austin
YPM	Yale Peabody Museum of Natural History, Yale University; New Haven, Connecticut.

Abbreviations Used in the Figures

art.par.	M. articuloparietalis
Ati.	atlantal intercentrum
Atn.	atlantal neural arch
Atp + axi.	fused atlantal pleurocentrum plus axial intercentrum
Ax.	axis vertebra
BO.	basioccipital
C.	caudal vertebra
D.	dentary
EO.	exoccipital
F.	femur
H.	humerus
int.art.	M. interarticulare
int.sp.	M. interspinalis
int.tr.	M. intertransversarii
l.cap.tran.cap.	M. longissimus capitis transversalis capitis
l.cap.tran.cer.	M. longissimus capitis transversalis cervicis
lev.cos.	M. levator costae
l.d.	M. longissimus dorsi
lig.nu.	ligamentum nuchae
M.	muscle
N.	nasal
obl.cap.inf.	M. obliquis capitis inferior
obl.cap.mag.	M. obliquis capitis magnus
obl.cap.sup.	M. obliquis capitis superior
Op.	opisthotic
P.	parietal
PP.	postparietal
pro.	process
Pt.	pterygoid
P2.	pelvic girdle
Q.	quadrate
Qj.	quadratojugal
R.	radius
Rcpp.	M. rectus capitis posterior profundus
Rcps.	M. rectus capitis posterior superficialis
S.	stapes
semisp.	M. semispinalis dorsi
semi.sp.cer.	M. semispinalis cervicis capitis
sp.cap.	M. spinalis capitis
sp.cer.	spinalis cervicis
Sp.d.	M. spinalis dorsi
Sq.	squamosal

Abbreviations

ST.	supratemporal
T.	tibia
U.	ulna
V.	vertebra, presacral

Abstract

The structure of the axial skeletons of a variety of Late Paleozoic tetrapods is described in detail. Although many aspects of their morphology are conservative, a conspicuous exception to the homogeneous vertebral structure is the presence of alternation of neural spine height and neural arch construction. Alternation of neural spine height, swollen neural arches, and tilted zygapophyses appear to belong to a functional complex common to a wide variety of the earliest truly terrestrial tetrapods. These include: seymouriamorph, diadectomorph, and certain microsaurian amphibians, almost all members of the reptilian family Captorhinidae; the better-known members of the araeoscelidian reptiles; and the pelycosaur *Varanosaurus acutirostris*.

The configuration of the zygapophyses indicates that vertebral dorsiflexion was an important component of axial movement in addition to lateral bending. The pattern of neural spine height alternation appears to have occurred usually near the limb girdles. It apparently permitted paired Mm. interspinales to skip over low-spined vertebrae and attach only to tall-spined vertebrae. M. spinalis dorsi and M. semispinalis dorsi were also attached to tall-spined vertebrae in an alternating fashion. It is hypothesized that this muscular configuration, along with the angulation of the zygapophyses, provided a mechanism for active columnar dorsiflexion. Such dorsiflexion would have aided in the recovery stroke of the step cycle by preventing outward rotation of the neural spines toward the columnar convexity during lateral bending. Further, such flexion could have provided a small amount of additional pressure to the contralateral limb performing the power stroke.

The phylogenetic significance of neural spine height alternation is difficult to assess.

1

INTRODUCTION

Most recent studies on the structure and function of bone have emphasized function in extant vertebrates. Functional studies do exist for fossil vertebrates, but the difficulty of the study often increases with the relative ages of the specimens. Extant forms do not always provide reasonable analogs for comparative study, and the often distorted condition of fossil material can be a serious deterrent.

Functional interpretations of the appendicular skeleton and associated muscular reconstructions are available for a number of Paleozoic tetrapods (e.g. Romer, 1922; Miner, 1925; Holmes, 1977, 1980), and cranial structure has been treated by more authors than can be listed here. The vertebral column has been used as a source of systematic characters for tetrapods (e.g. Romer, 1947, 1956, 1966), and several authors have detailed the vertebral column within the context of larger systematic descriptions (e.g. White, 1939; Bystrow, 1944; Carroll, 1968, 1970). However, fewer functional interpretations exist for the vertebral column and its associated axial musculature. Panchen (1967, 1977) and Parrington (1967) have provided functional interpretations of early tetrapod vertebral columns. Laerm (1979) and Lauder (1980) have emphasized the intersegmental position of neural arches, a location functionally important for communication of muscular action to the bony vertebral column. Olson (1936) recognized the importance of integrating studies of myology and osteology and produced what remains the definitive work on the evolution of the early tetrapod vertebral column and its associated musculature in Late Paleozoic tetrapods. However, new and better prepared material is now available, and a reconsideration of the structure, function, and importance of the vertebral column in advanced Paleozoic amphibians and primitive reptiles may now be proposed.

Among Late Paleozoic tetrapods, little variation has been recognized between the vertebral columns of different taxa, or between the individual vertebrae of a single animal. However, Carroll (1968) and, later, Vaughn (1970) noticed that such homogeneity was not always the rule. A number of Late Pennsylvanian and Early Permian tetrapods display a marked alternation of tall, well developed neural spines with lower, less highly developed neural spines. A concomitant variability in the structure of the supporting neural arches often exists as well. Many early authors ignored this phenomenon, attributing it to imperfect preservation (e.g. Romer, 1946); but in fact, a number of Permo-Carboniferous tetrapods, from a wide variety of taxonomic groups,

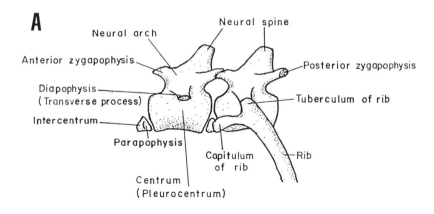

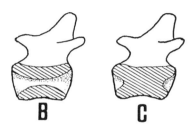

Figure 1. General vertebral structure in Permo-Carboniferous tetrapods. A, features of two generalized tetrapod vertebrae. B, sagittal section of a notochordal vertebra. C, sagittal section of an amphicoelous vertebra. Anterior is to the left in all portions of the figure. Hatching indicates sections through solid bone.

display such an alternation of neural spine height. Because of its widespread occurrence, and its radical departure from the basic vertebral anatomy that was previously accepted in most of the forms that possess it, a morphological and functional interpretation is presented.

Only Vaughn (1970) and Carroll and Gaskill (1978) have considered the function of alternation in height of the neural spines in any way. Vaughn's (1970) report on alternation was brief, and he later admitted that it was not sufficient to explain the phenomenon in all of the animals that displayed it (Vaughn, 1972). Carroll and Gaskill (1978) restricted their interpretation to a few lines of comment. The present study represents an attempt to remedy the currently incomplete perceptions of vertebral function in Paleozoic tetrapods.

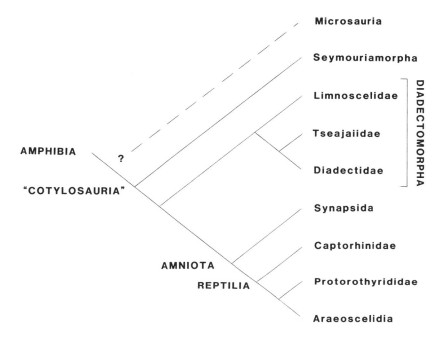

Figure 2. Hypothesis of relationships of taxa included in this study. Based primarily on Heaton (1980), Heaton and Reisz (1986), Gauthier et al. (1988), and Panchen and Smithson (1988).

GENERALIZED VERTEBRAL STRUCTURE IN TETRAPODS

The tetrapod vertebral column can be divided into presacral and caudal regions, separated by the sacral vertebrae specialized for support and attachment of the pelvic girdle. Romer (1956) admitted to difficulty in a precise definition of sacral vertebrae. Although there is acknowledged variability in this characteristic (e.g. Bystrow, 1944), they are defined as those vertebrae that, with associated ribs, support the ilium through direct ilio-sacral contact. The two anteriormost vertebrae are the specialized atlas and axis vertebrae. The remaining presacrals may be considered dorsals, and they display a gradual change in most, but not all, features from anterior to posterior. A diagrammatic illustration of basic vertebral structures is provided in Figure 1.

There have been some attempts to define a cervical region in primitive tetrapods, but interpretations vary. Romer (1956) defined cervical vertebrae as those anterior vertebrae whose ribs did not reach the sternum. The fact that most sternal ribs in reptiles and amphibians make contact with the sternum via cartilagenous extensions makes such a definition difficult to use with Paleozoic tetrapods. Heaton and Reisz (1980) defined the cervical region of captorhinid reptiles as that region with a less conspicuous swelling of the neural arch, a greater degree of zygapophyseal tilt, and attachment of expanded holocephalous ribs. Expanded ribs are found fairly far posteriorly, however, as they help to support the pectoral girdle via the M. serratus ventralis, so this later definition is also difficult to apply. Except for the specialized atlas and axis and cases in which a distinct cervical region can be noted as in *Araeoscelis* (Vaughn, 1955; Reisz, et al., 1984) or *Petrolacosaurus* (Peabody, 1952; Reisz, 1980), the presacral vertebrae is not subdivided regionally.

Caudal vertebrae are those posterior to the sacrals. Estimates of the length of the caudal region vary from about 40 for seymouriamorph amphibians and captorhinid reptiles (White, 1939; Fox and Bowman, 1966) to 60 in small captorhinids (Heaton and Reisz, 1980) and 50 to 70 in other primitive reptiles (Romer, 1956).

METHODS

Structural Context

As alternation in neural spine height and structure is a poorly documented phenomenon, a thorough osteological description of vertebral structures in the forms that exhibit it is a primary objective. As this condition is not known to occur in any extant tetrapod, use of extant organisms as a basis of comparison or as models of morphological interpretations is difficult. Initial descriptions of the reptilian family Captorhinidae will provide a detailed basis for comparison with other groups known to display the phenomenon. Muscular reconstructions, however, must have some theoretical basis. In all cases of proposed muscular reconstructions, a conservative approach has been taken. The structure of those muscles that appear to have been similar to those of extant forms was determined first, and only then were more unique structures restored relative to them. An exact comparison remains impossible, as the alternation appears to have been restricted to Late Pennsylvanian and Early Permian tetrapods. The modern lizard *Iguana iguana* and the crocodilian *Caiman* were dissected and used as a basis of comparison whenever possible. Further comparisons were made to literature descriptions of axial musculature for *Iguana* (Olson, 1936), *Sphenodon* (Osawa, 1898), and other extant amphibians and reptiles (Nishi, 1916). Once a hypothetical arrangement of musculature was proposed, confirmation of such an arrangement was sought in a histological examination of the abundant vertebral material available for the captorhinid reptile *Captorhinus aguti* from the fissure fills of the Richard's Spur locality (Lower Permian, Clear Fork Group Equivalent) of Oklahoma. Once this model has been established, a survey of other groups that exhibit alternation will be possible and a general morphological and functional model for the phenomenon may be proposed.

Phylogenetic Context

A historical review of tetrapod classification and relationships is neither within the range, nor is the purpose, of this study. Fortunately, a hypothesis of phylogenetic relationships that encompass the taxa reviewed here has recently been provided by the works of Heaton (1980), Heaton and Reisz (1986), Gauthier et al. (1988), and Panchen and Smithson (1988). An eclectic (and necessarily somewhat simplified) summary of their works is provided in Figure 2. Not all groups treated in their analyses are included in Figure 2, but it does serve to supply a context within which the described axial characteristics may be discussed.

2

THE FAMILY CAPTORHINIDAE

The reptilian family Captorhinidae has been central to the morphological and phylogenetic interpretations of most primitive reptiles, and it has been the subject of a succession of systematic reviews (Case, 1911; Seltin, 1959; Romer, 1966; Gaffney and McKenna, 1979; Ricqles, 1984). The majority of the literature on the Captorhinidae has been concerned with cranial structures, and function has not been a primary concern. No comprehensive examination of the captorhinid vertebral column has been done to date, and with the availability of important new materials, such an analysis will form a basis of comparison for this study.

Captorhinus, Eocaptorhinus, and *Labidosaurus* are the most thoroughly studied members of the Captorhinidae. The vertebral column in *Labidosaurus* has been redescribed recently (Sumida, 1987), as has that of *Eocaptorhinus* (Dilkes and Reisz, 1986), but a complete description of the column in *Captorhinus* has been notably absent.

The first definition of the family Captorhinidae was provided by Case (1911) based on the genus *Captorhinus* (Cope, 1882, 1886), but the characters and specimens were of limited utility. Fox and Bowman (1966) discussed the osteology and relationships of *Captorhinus*, although their study utilized primarily disarticulated elements, many of which are now assignable to the genus *Eocaptorhinus* (Heaton, 1979). Gaffney and McKenna (1979) and Heaton (1979) redefined the family Captorhinidae and included as its most primitive member the genus *Romeria*, previously assigned to the Protorothyrididae (= Romeriidae; Price, 1937; Clark and Carroll, 1973).

Well-preserved and essentially complete vertebral materials are available for *Captorhinus, Eocaptorhinus,* and *Labidosaurus.* The vertebral column is less completely represented in the genera *Protocaptorhinus, Captorhinikos,* and *Labidosaurikos,* but enough material is available to assess the presence of alternation of neural spine height and structure. Alternation is not known to occur in *Romeria*, the most primitive genus of the Captorhinidae (Gaffney and McKenna, 1979); however, only a small portion of the column is known and the possible presence of the phenomenon cannot be ruled out at this time. The entire column is preserved in *Rhiodenticulatus heatoni* (Berman and Reisz, 1986), but it is exposed in ventral aspect only, precluding examination of patterns of the neural spines. The skull of the large African form *Moradisaurus grandis* has been described in detail (Taquet, 1969; Ricqles and Taquet, 1982), but no

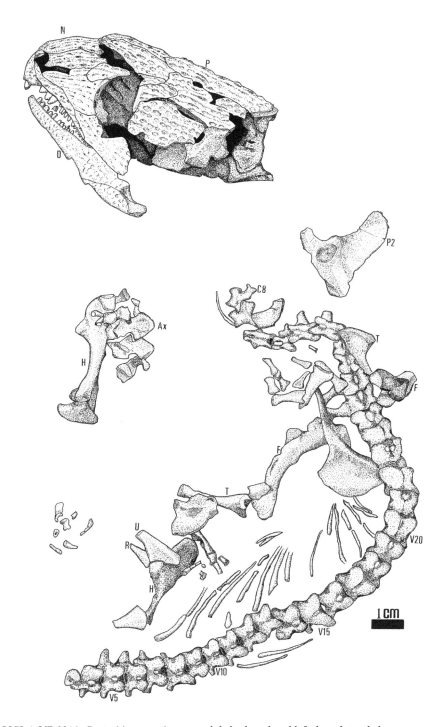

Figure 3. UCLA VP 3214, *Captorhinus aguti*, seen mainly in dorsal and left dorsolateral view.

description of the postcranium has yet been provided, and the specimen was not available to this author. No postcranial materials are known for the large Upper Permian *Hectagomphius kavejevi* (Vjushkov and Chudinov, 1957; Olson, 1962) of the Soviet Union. Vertebral materials are known for the large Upper Permian forms of North America, *Kahneria seltina* (Olson, 1962) and *Rothianiscus multidonta* (Olson and Beerbower, 1953; Olson, 1962). While the variability in what remains of the neural arches and spines in these specimens is suggestive of the possibility of alternation of neural spine height, their extraordinarily poor preservation does not allow a confident description of axial structures.

Despite the lack of diagnostic axial material for certain of the captorhinids, they still provide the largest sample of material in which alternation of neural spine structure is manifested. For this reason they represent the initial descriptions presented in this study.

The genera *Captorhinus* and *Eocaptorhinus* are closely related (Heaton, 1979), distinguished only by differences in the number of maxillary and dentary tooth rows. The validity of generic distinction between the two has been debated (Bolt and DeMar, 1975; Heaton, 1979; Heaton and Reisz, 1980), but they are treated as distinct taxa here in line with the arguments of Heaton and Reisz (1980). Axial structure in the two taxa are similar in many respects, and separate descriptions would entail a great deal of repetition. Therefore, the vertebral column of *Captorhinus* is described in detail and that of *Eocaptorhinus* follows, with only the important differences between the two noted.

CAPTORHINUS AGUTI

Most specimens of *C. aguti* possess 25 presacral and 2 sacral vertebrae. Fox and Bowman (1966) based this estimate on a number of almost complete columns and evaluations drawn from a large amount of disarticulated material. AMNH 4434 and uncatalogued AMNH material confirm this number with complete presacral columns, but UCLA VP 3214 (Figure 3) possesses 26 presacral vertebrae. Fox and Bowman estimated the number of caudal vertebrae to be 36; however, Heaton and Reisz (1980) regard 60 as a more probable minimum.

Atlas-Axis Complex. The atlantal intercentrum (Figure 4) is worn, but visible in UCLA VP 3214. It is wedge-shaped and more blocky in appearance than the intercentra of the dorsal vertebrae. The proatlas was tentatively identified by Fox and Bowman (1966), and is preserved in UCLA VP 3214. Fox and Bowman described it as consisting of "a broad basal plate that possesses a delicate posteriorly directed spine on its dorsal surface. Posteriorly a broad edge contacted the atlantal arch. Anteriorly the halves of the proatlas adjoined paired facets of the exoccipitals bordering the foramen magnum." The posteriorly directed spine may be somewhat blunter than described above. The atlantal neural arch bears a well-developed footplate and a slightly constricted "neck" leading to a posteriorly directed spine.

The axial intercentrum and atlantal centrum are fused, resulting in a ventrally keeled, subtriangular disc (Peabody, 1952: Fig. 10A), pierced for passage of the notochord. The rib articulation is directed posteriorly at an angle of about 30 degrees.

The axis of *C. aguti* is immediately recognizable by its keeled centrum and massive neural spine. In some specimens the keel angles slightly anteroventrally. The posterior face of the axial centrum is only lightly bevelled for reception of an intercentrum. The neural arch is not 'swollen' as are those of more posterior vertebrae.

The axial neural spine is subtrapezoidal in lateral view and extends beyond the anterior and posterior limits of the neural arch. The spine is highest near its posteriormost level and slopes anteroventrally toward the atlas at an angle of about 20-25 degrees. A thickened ridge runs the length of its dorsolateral aspect.

The axial neural spine is narrowly triangular in dorsal aspect (Figure 4B). A narrow furrow that terminates at the midlength of its dorsal margin separates posteriorly into a pair of processes. Each of these processes is further subdivided by a small groove that continues down the posterior margin of the spine, resulting in a total of 4 nipple-shaped processes.

Examination of well-preserved material from the fissure fills of the Fort Sill locality in Oklahoma, and comparison with structures determined for *Eocaptorhinus* by Heaton (1979) and the living *Iguana* (pers. obsv.; Nishi, 1916; Olson, 1936), allows reasonably confident description of the surfaces of muscular attachments of the atlas-axis complex (Figures 4B and C, 5, and 6). The largest of the occipital muscles attaching to the axis was the M. obliqus capitis magnus, which had a somewhat rectangular-shaped origin on the lower two-thirds of the axial neural spine and inserted along the posterior surface of the paroccipital process on the opisthotic bone. Just dorsal to the origin of the M. obliqus capitis magnus, the M. rectus capitis posterior had a longer and distinctly narrower origin. Its two subdivisions occupied a more medial position than the M. obliqus capitis magnus. The M. rectus capitis posterior superficialis inserted on the concave posteromedial portion of the postparietal, whereas the M. rectus capitis posterior profundus inserted at a more anterior position on the posteromedial aspect of the supraoccipital.

The M. spinalis capitis originated from the lateral aspect of the pronounced dorsal lip that runs the length of the axial neural spine and the taller spines of the anteriormost presacral vertebrae. It inserted dorsal to the insertion of the M. rectus capitis posterior on the postparietal.

In well-preserved specimens small scars suggest that the M. obliqus capitis superior may have originated at the anteroventral limit of the axial neural spine, and possibly at the posteriorly projecting spine of the atlantal neural arch. It inserted on the paroccipital process of the opisthotic lateral to the M. obliqus capitis magnus. Although there is no definitive evidence for an insertion of the M. obliqus capitis inferior independent of that of the obliqus capitis superior, there appears to be an indication of its origin at the most anterodorsal portion of the atlantal neural arch.

Mm. interarticulares connected the posterior zygapophyses of the atlas and axis. More dorsally, paired interspinous muscles probably took their attachment in the two lateral furrows on the posterior face of the axis. The median furrow probably accommodated a supraspinous ligament. This may well have been a functional extension of the nuchal ligament that attached to the large anterior projection of the axial neural spine. The cranial attachment of the nuchal ligament was at a pronounced median ridge

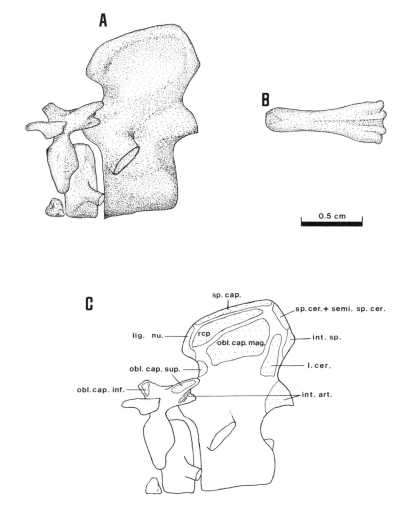

Figure 4. Atlas-axis complex of *Captorhinus aguti*. A, reconstruction of the atlas-axis complex, left lateral aspect. B, axial neural spine, dorsal aspect. C, areas of muscle attachment on the atlas-axis complex. Based on UCLA VP 3214 and uncatalogued material in the collections of the Field Museum of Natural History.

of the supraoccipital. This interpretation of an extended nuchal ligament is somewhat different from that of Vaughn (1970).

Dorsal vertebrae. The shape of the dorsal centra in end view is that of a squat oval. Both faces are bevelled ventrally for reception of intercentra, the anterior face to a greater degree. Proceeding posteriorly through the column there is a gradual lengthening of the centrum; the anteriormost centra being about 80-85% the length of posterior

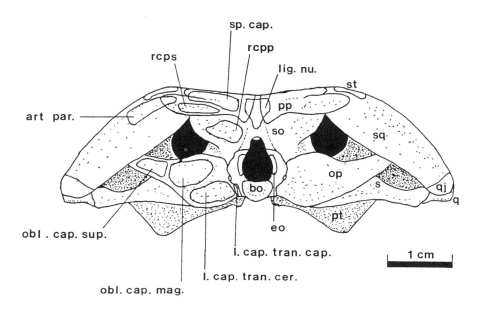

Figure 5. Areas of muscle attachment on the occipital aspect of the skull of *Captorhinus aguti*. Based primarily on UCLA VP 3214, UCLA VP 754, UCLA VP 3752, UCLA VP 3753, KUVP 9978, and information from Heaton (1979).

centra. The first two dorsals possess a midventral keel in line with those of the atlas and axis. Posterior to the 3rd dorsal (5th presacral) the centrum is slightly pinched at its midpoint, but by midcolumn the centra are smoothly spool-shaped. In sagittal section the centrum is hourglass-shaped.

Anteriorly the transverse processes in *Captorhinus* are very well developed, flaring laterally to a flattened articulation with the rib head (Figures 7 and 8); they extend out as much as two-thirds the width of the centrum. The articular face of the transverse process has a dumbbell-shaped outline, with a thin bony web between the diapophysis and parapophysis. The axis of this connection slants back 20-25 degrees from the vertical axis (Figure 8), and the articular face is angled downward approximately 30 degrees.

By the middle of the column, the articular faces of the transverse processes angle posteriorly as much as 45 degrees. They face laterally, the articular surface is narrower, and the parapophysis becomes smaller than the diapophysis, barely making contact with the anterior limit of the centrum. More posteriorly the transverse processes fade, and the rib articulations are restricted to a diapophysis by the 20th presacral. As the ribs decrease in size and as contact with the intercentrum is lost, the diapophysis becomes positioned closer to the anterior edge of the pedicel (Figure 11).

In well-preserved specimens, a dorsal ridge on the transverse process indicates a portion of the M. longissimus dorsi insertion (Figure 9). This insertion continues onto

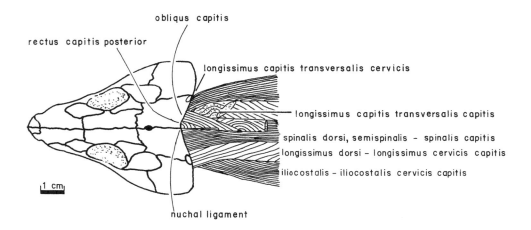

Figure 6. Reconstruction of occipital and anterior epaxial musculature of *Captorhinus aguti*.

the dorsal margin of the proximal portion of the rib. The M. longissimus dorsi continued anteriorly as the M. articuloparietalis, inserting on the postparietal and the medial portion of the squamosal (Figure 5). By comparison with *Iguana* (in a manner similar to determination of occipital musculature) and *Sphenodon* (Osawa, 1898), Mm. levator costae would have originated from the posterodistal edges of the transverse processes. Deep to these, the Mm. intertransversarii ran from the posterior to anterior edges of successive transverse processes. The more anterior attachment was larger, whereas the posterior was limited to a cleft between the transverse process and the anterior zygapophysis.

The pedicels supporting the neural arch are attached toward the anterior portion of the centrum. They become stouter toward the rear of the column with those near the end of the column being 60% longer than the anteriormost pedicels (Figure 11). As the neural arches become more "swollen" toward the posterior portion of the column, the orientation of the pedicels changes from vertical to a lateral tilt of 7-10 degrees.

In the anterior portion of the vertebral column the outline of the neural canal is a flattened, curved oval, about twice as wide as it is high. More posteriorly the difference between height and width decreases, assuming a ratio of 2:3 by presacral 18. The floor of the neural canal is smooth, the lateral walls have a slight medial constriction, and the roof displays minute pitting in some specimens.

Although most authors (Fox and Bowman, 1966; Romer, 1956, 1966) have described the zygapophyses of *Captorhinus* as essentially horizontal in orientation, there is a slight zygapophyseal tilt present throughout the column. The zygapophyses are angled 7-10 degrees to the horizontal in the anteriormost vertebrae. Zygapophyses of the more posterior portion of the column tilt at a slightly greater angle, and the articulations approach a cup-shaped form (Figure 7).

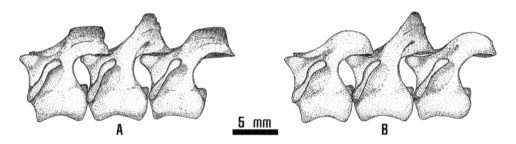

Figure 7. Reconstruction of dorsal vertebrae of *Captorhinus aguti* showing different expressions of alternation of neural spine height and structure. A, alternation of neural spine height and structure with low-spined vertebrae possessing narrow, wedge-shaped neural spines. B, alternation of neural spine height and structure with low-spined vertebrae possessing no clearly defined neural spine. A is based on UCLA VP 754, UCLA VP 3176 and uncatalogued materials in the collections of the Field Museum of Natural History. B is based on UCLA VP 1735 and uncatalogued materials in the collections of the Field Museum of Natural History.

The configurations of the neural arches and spines in *Captorhinus* show both serial as well as more localized individual variations. The arches are not swollen anteriorly, but gradually increase in width until they are approximately 25% wider by the posterior termination of the dorsal series.

Vaughn (1970) first reported alternation of neural spine height for *Captorhinus*, but the study was limited to short strings of vertebrae. Articulated columns now allow a more thorough description of this phenomenon. Alternation of spine height is found in at least three different patterns in specimens attributable to *Captorhinus*.

Where alternation occurs, tall conical spines alternate with lower spines. The lower spines are usually long, narrow, anteroposteriorly extended ridges, but often no spine is present at all and there exists only a rounded-off neural arch (Figures 7 and 8). The tall neural spines come to a pointed apex. The associated neural arch sweeps up steeply, accentuating its difference from the more rounded neural arches of low-spined types, and exhibit paired scars for attachment of the Mm. interspinales anteriorly and posteriorly (Figures 8 and 9). Low-type spines show no such scars of attachment, and when present rarely reach the level of the medial axis of the paired Mm. interspinales.

In regions of the column lacking alternation the tip of the neural spine is heavier and more rugose, indicating what may have been a firm attachment for a supraspinous ligament. Scars of attachment for Mm. interspinales are present, but not as extensive as in regions of alternation. In well-preserved specimens, prominent ridges of attachment for the M. spinalis dorsi and M. semispinalis are visible along the posterolateral face of the neural spine and anteriorly along the neural arch. In areas of alternation of spine height, there is no obvious scar or ridge to indicate attachment for these muscles on either tall or low types, although the taller neural spines probably provided some attachment for the muscle. Anteriorly, the associated M. spinalis cervicis and M. semispinalis cervicis passed to ridges on the posterior margin of the axis and the 3rd presacral. Although alternation is sometimes present in this region of the column, the 1st dorsal (3rd presacral) is always a tall type, which could provide such an attachment.

Attachments of the Mm. interarticulares are present on vertebrae conforming to both alternating and non-alternating patterns. They originated from anterolaterally facing depressions just in front of the posterior zygapophyses. Their presumed insertion posterior to the anterior zygapophyses and the posterior zygapophyses of the next anterior vertebra is rarely seen. The pattern of muscular attachments is distinctly different in regions of the column that exhibit alternation and those that do not. In regions of alternation, Mm. interspinales, M. spinalis dorsi, and M. semispinalis dorsi attach to the neural spines and arches of alternate vertebrae, whereas there is no skipping of vertebrae in regions where alternation is not expressed. However, in both patterns of vertebral structure, attachments for Mm. interarcuales cannot be differentiated from those of the interspinous muscles.

Two distinct patterns of alternation can be seen in articulated columns of *Captorhinus*. Most complete columns show alternation of neural spine height in the anterior (presacrals 3 through 7) and posterior dorsal (20 through 25) vertebrae, but not in the mid-dorsals (Figure 11). AMNH 4332 and AMNH 5494 indicate that 18 may also be a low type. UCLA VP 3214 shows alternation only in the rear of the column, a pattern similar to that described for *Eocaptorhinus* (Heaton and Reisz, 1980). This specimen possesses multiple tooth rows and is definitely assignable to the genus *Captorhinus*.

UCLA VP 3214 (Figure 3) is unique in that it possesses 26 presacral vertebrae. All other captorhinids whose presacral numbers are known possess 25 (Fox and Bowman, 1966; Heaton and Reisz, 1980; Sumida, 1987). Such variability is not surprising, as many reptilian genera show variation in vertebral numbers (Romer, 1956). In this specimen, 14 is low, and alternation proceeds rearward to the last presacral. There is one interruption in the pattern: both 19 and 20 are tall types. Such interruption in the pattern is not uncommon in other taxa (Vaughn, 1972; Sumida, 1987).

The two patterns described above are the only ones recognizable from articulated specimens. Uncatalogued materials at the Field Museum of Natural History show alternation of neural spine height in short strings of vertebrae from the anterior and middle portions of the presacral column. Thus, it is conceivable that alternation of spine height may have occurred throughout the column as well. Dilkes and Reisz (1986) report this condition for two specimens of *Eocaptorhinus*. The Field Museum materials are from the Fort Sill, Oklahoma, locality (see Appendix for details), and Heaton (1979) estimates that only one in 20 captorhinid specimens from Fort Sill is *Eocaptorhinus*, so *Captorhinus* is probably represented in these materials. *Captorhinus* may display at least three different patterns of neural spine height alternation, but the specific varieties of patterns probably are not as important as the consistency with which the pattern occurs near the limb girdles.

In order to understand the muscular structures associated with alternation of neural spine and neural arch construction more completely, histological sections were prepared for both high- and low-type neural spines (Plate 1). Sectioned specimens were limited to those with undamaged neural spines.

Frontal sections made at various levels of both tall- and low- type neural spines revealed variations in thickness of compact lamellar bone in different regions of

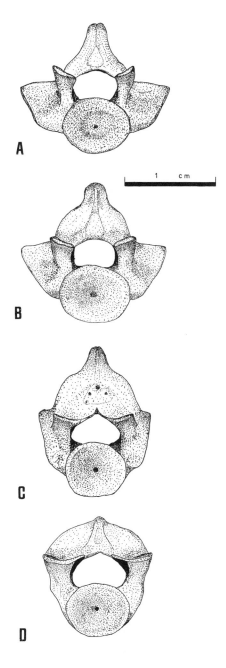

Figure 8. Dorsal vertebrae of *Captorhinus aguti* in anterior view. A, 3rd presacral vertebra. B, 7th presacral. C, 15th presacral. D, 21st presacral, with low-type neural spine. Based on AMNH 4332, AMNH 4334, AMNH 5494, UCLA VP 1735, UCLA VP 3214, and uncatalogued material in the American Museum of Natural History and the Field Museum of Natural History.

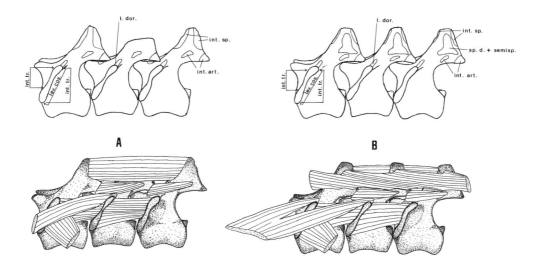

Figure 9. Areas of muscle attachment on dorsal vertebrae and reconstructions of muscles attaching to the dorsal vertebrae in *Captorhinus aguti*. A, muscle attachments and reconstruction of muscles in vertebrae with alternation of neural spine height. B, muscle attachments and reconstruction of muscles in vertebrae that do not exhibit alternation of neural spine height. Based on the same specimens as Figure 8.

individual spines, as well as between different types of spines. The anterior wall of tall neural spines is almost always thicker than either the lateral or posterior walls, often by as much as 33%. The differences in thickness are consistent through the length of the neural spine except at the very tip, where all of the lamellar walls are quite thin. Mm. interspinales may have swept up to insert on a supraspinous ligament as well as the posterior face of the neural spine, eliminating the need for a thick-walled attachment; or the posterior attachment may have been spread over a larger area, with the compensating difference reflected in a thinner posterior wall. However, the extremely thin walls of the dorsal tips of tall-type neural spines in regions of alternation seem to preclude the possibility of a very heavy supraspinous ligament.

Fewer frontal sections of low-type neural spines were available for study, as they make up only about 20% of a typical *Captorhinus* presacral column. Those that are available exhibit disproportionately thicker lateral walls than those of tall-type neural spines. The plane of section could not travel normal to the surface of the neural spine, giving it a disproportionately thicker appearance. Even with this distortion taken into consideration, the anterior and posterior walls of the low-type spines are conspicuously thinner than the same walls of tall-type spines, as well as relative to their own associated lateral walls. A proliferation of branching and anastomosing vascular canals is seen in the lateral walls of the low-type spines. Enlow and Brown (1957) indicated that the vascular canals found in *Captorhinus* usually possessed a simple longitudinal orientation. Vascular canals are only rarely found in tall-type neural spines but, when detected, they are usually directed toward the anterior tip and posterolateral sides of the spine. Low-type spines exhibit vascular canals much more frequently, and they exhibit a complicated branching pattern.

Enlow (1969) noted that in the long bones of *Captorhinus* the medulla lacks cancellous trabeculae. Similarly, most low-type spines have no supporting trabeculae. The smaller and narrower the spine, the less room there is for trabecular structures, but their absence is noteworthy nonetheless. When trabeculae are found in low-type neural spines they almost always have a transverse orientation, acting essentially as crossbraces to the narrow, but anteroposteriorly elongated, structure.

Trabeculae are much more common in tall-type neural spines. Anteriorly placed trabeculae in tall-type neural spines have a consistent orientation toward, and connection to, the anteromedial lamellar wall. More posteriorly, there is a consistent orientation of trabeculae toward the posterolateral wall of the neural spine and just opposite the presumed attachment of the M. spinalis dorsi and M. semispinalis dorsi.

Sharpey's fibers may be observed within the walls of some specimens of tall-type neural spines. They are directed forward in the anterior surface of the neural spine. Laterally, they follow essentially the same line as that projected by the posterolaterally directed trabeculae. In these and other specimens of tall neural spines, the orientations of the lacunae are similarly aligned. Further, when vascular canals are observed in tall-type neural spines, they are likewise oriented. The orientation of lacunae in low spines simply parallels the concentric layers of compact bone and does not indicate potential muscular attachments in any way. No evidence of Sharpey's fibers has been found in any of the sectioned specimens of low-type neural spines.

Sacral vertebrae and ribs. Captorhinus possessed two sacral vertebrae (Figure 10). The centrum of the first sacral is fused to that of the last dorsal, and apparently little or no movement was possible between them. The second sacral centrum was slightly shorter and lighter. Its articulation with the first sacral vertebra is also quite firm. No ventral ridges are visible on the materials available for study.

Laterally, the sacral centra and pedicels are dominated by large articulations for the sacral ribs. That of the first rib is trapezoidal in outline shape, taking up at least two-thirds the length of the centrum and angling downward approximately 30-35 degrees. The second costal articulation is not as extensive; it angles forward just slightly, directing the second sacral rib to its contact with the first.

The neural arches of the sacral vertebrae are considerably narrower than those of presacral vertebrae, rising to the neural spines much more steeply. The anterior and posterior zygapophyses are horizontal in orientation. Sacral neural spines do not contribute to the pattern of alternation in height. They are blocky in lateral aspect, but laterally compressed in dorsal view. Their dorsal extent is greater than that of either the posterior dorsals or the anteriormost caudals. The tips of the spines are quite rugose, and the axes of the spines tilt slightly forward (Figure 10B).

The first sacral rib is much larger than the second, accounting for the majority of the curved ilial articulation (Figure 10B), and has a very constricted, well-defined neck. The second sacral rib angles slightly anteriad to attach to a shallow depression on the underside of the posterodistal edge of the first. There is no evidence of a large last

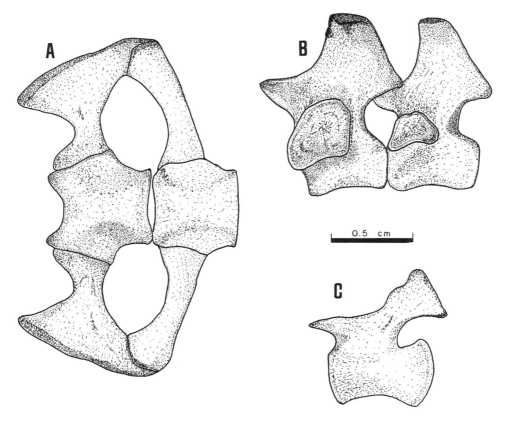

Figure 10. Sacral vertebrae, sacral ribs and caudal vertebra of *Captorhinus aguti*. A, sacral vertebrae and ribs, ventral aspect; anterior is to the left. B, sacral vertebrae, left lateral aspect. C, 8th caudal vertebra, left lateral aspect. Based on UCLA VP 3214.

presacral rib contributing to the support of the ilium via a ligamentous connection as in *Labidosaurus* (Sumida, 1987).

The Mm. spinalis dorsi and semispinalis, and M. longissimus dorsi apparently extended without interruption over the sacral region into the M. extensor caudae medialis and M. extensor caudae lateralis, respectively. The M. longissimus dorsi almost surely took at least part of its origin from the first sacral rib and probably from the second as well. Mm. interspinales were probably much more tendinous in this region, and may have have been interwoven with a well-developed supraspinous ligament. A similar arrangement is seen in modern iguanid lizards. These muscular patterns are also similar to those proposed for a variety of other early Permian tetrapods (Olson, 1936), which is not surprising, considering the conservative nature of the structures in this region (Olson, 1936).

Caudal vertebrae and ribs. The first caudal centrum is very much like that of the second sacral. The following centra decrease in length extremely gradually. The first four caudals bear short, recurved ribs; the next two bear very short, fused ribs; and those remaining bear none. The width of the neural arches decreases markedly and the neural spines decrease in height posteriorly. The plane of articulation between the anterior and posterior zygapophyses is tilted as much as 20 degrees to the horizontal plane in the caudal region. Presumably the most posterior caudals were reduced to simple cylindrical rods. The materials examined showed no evidence of transverse cartilagenous septa that could be associated with caudal autotomy. However, Price (1940) recorded the presence of such septa in certain specimens attributed to *Captorhinus*.

The anteriormost caudals show patterns of muscular attachments similar to those of the Mm. spinalis dorsi and semispinalis in regions of the dorsal column in which there is no alternation of neural spine height. These two muscles probably contributed equally to the M. extensor caudae medialis.

Intercentra. Intercentra were present through the presacrals, the sacrals, and to approximately the 6th caudal. Farther posteriorly they are represented by posteroventrally directed chevrons. The intercentra are thin, crescentic in form, and smaller in the sacral and caudal regions than those of the presacral series.

Presacral ribs. Atlantal ribs are not preserved in any of the materials at hand, but must have been very short. The axial ribs were short, relatively broad, and slightly recurved. The next 4 ribs are broad, spatulate in shape, and also slightly recurved. They probably functioned in support of the pectoral girdle via a strong M. serratus ventralis (Holmes, 1977).

The capitulum and tuberculum of the rib head are connected by a thin bony web. A bony connection also angles down from the capitulum to the shaft, giving the head a triangular shape in anterior view, a characteristic lost as the parapophysis and the associated capitulum decrease in size posteriorly. Distal to the head, the rib is constricted into a thin, cylindrical rod. Costal length increases gradually about midcolumn, then decreases rapidly. The last few dorsal vertebrae have very short ribs.

At the proximal end of well-preserved ribs a dorsally facing, oblique ridge may be seen. It continues up onto the transverse process and apparently served as the insertion for the M. longissimus dorsi. More distally, the M. iliocostalis dorsi inserted along the dorsolateral edge of the most sharply curved portion of the rib shaft. Presumably, external and internal intercostal musculature was attached more deeply and distally, but the limits of the origins and insertions of these muscles cannot be determined from the available materials.

EOCAPTORHINUS LATICEPS

The proportions of the individual vertebrae, and the relative proportions of tall- and low-type vertebrae in *Eocaptorhinus*, are similar to those in *Captorhinus*. All known material in *Eocaptorhinus* indicates a presacral count of 25 (Heaton and Reisz, 1980; Dilkes and Reisz, 1986). *Eocaptorhinus* possessed two sacral vertebrae and probably had about 60 caudals (Heaton and Reisz, 1980).

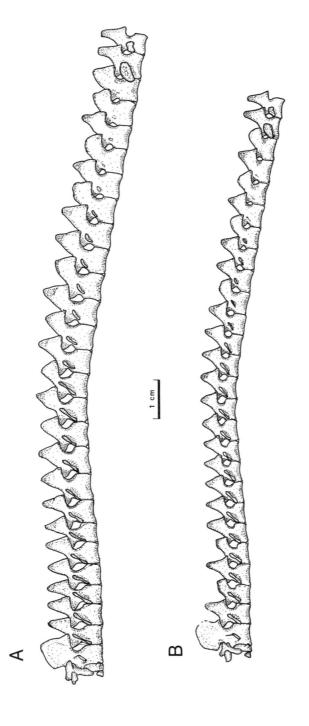

Figure 11. Reconstructions of the vertebral column in *Eocaptorhinus laticeps* and *Captorhinus aguti*. A is the form most commonly found in *E. laticeps*; B is the form most commonly found in *C. aguti*.

Specimens available for this study were not preserved well enough to permit a confident interpretation of occipital muscular attachments. Based on the work of Heaton (1979), the cranial attachments of the epaxial musculature of *Eocaptorhinus* would appear to be very similar to those proposed for *Captorhinus* above (Figure 6).

As in *Captorhinus*, regional distinction of the presacral column in *Eocaptorhinus* is difficult. Dilkes and Reisz (1986) report no ventral keel in the cervical vertebrae of *Eocaptorhinus*; but unlike *Captorhinus*, the ventral edges of the centra possess paired lips, and more posterior vertebrae do have paired, parallel ridges on the ventral aspect of the centrum. The proportions and structure of the centra, pedicels, and zygapophyses are similar to those in *Captorhinus*, but the neural arches and spines are slightly different. Anteriorly directed mammillary processes project from the lateral aspect of neural arches 6 to 13 (Heaton and Reisz, 1980; Dilkes and Reisz, 1986); presumably they were strengthened points of attachment for the M. spinalis dorsi and M. semispinalis dorsi.

Heaton and Reisz (1980) reported alternation of neural spine height in only the posterior portion of the presacral column of *Eocaptorhinus*. They did note that there was often an irregular, low-type spine at various positions in the anterior portion of the column. Dilkes and Reisz (1986) described alternation extending anteriorly through the middorsal region of *Eocaptorhinus*. The alternation in neural spine height is extremely subtle except in the last two vertebrae of the specimen (OUSM 10520B) described by Dilkes and Reisz (1986). No pattern similar to the concentration of alternation in the anterior and posterior regions of the presacral column in *Captorhinus* is visible in *Eocaptorhinus*. The pattern described for *Eocaptorhinus* by Heaton and Reisz (1980) is also found in at least one specimen of *Captorhinus* (UCLA VP 3214). A variety of patterns of alternation of neural spine height probably existed in both *Captorhinus* and *Eocaptorhinus*. It seems likely, however, that *Eocaptorhinus* did not show the predominant pattern seen in *Captorhinus*. It may be significant that both genera display the phenomenon near the pelvic girdle.

The neural spines of sacral vertebrae do not show the forward tilt seen in *Captorhinus*; otherwise, the sacral vertebrae and ribs, caudal vertebrae, intercentra, and ribs in *Eocaptorhinus* are essentially similar to those in *Captorhinus*.

LABIDOSAURUS

Of the primitive captorhinids, *Labidosaurus* (Figures 12 and 13) is among the best-represented by complete specimens. Early studies of *Labidosaurus* had a fairly narrow focus, as most authors concentrated almost exclusively on cranial materials (e.g. Williston, 1910; Cope, 1896; Huene, 1913). Elements of the postcranial skeleton have been described in more detail recently (Sumida, 1987, 1989a). The following description is in great part abridged from that work.

Labidosaurus has been treated by most authors in a general manner, usually regarded as similar to *Captorhinus* (Broili, 1904; Case, 1911; Olson, 1937; Romer, 1956; Williston, 1917). Whereas *Labidosaurus* does show some similarities to *Captorhinus*, the two are not identical, and the pattern shown by its vertebrae is far from repre-

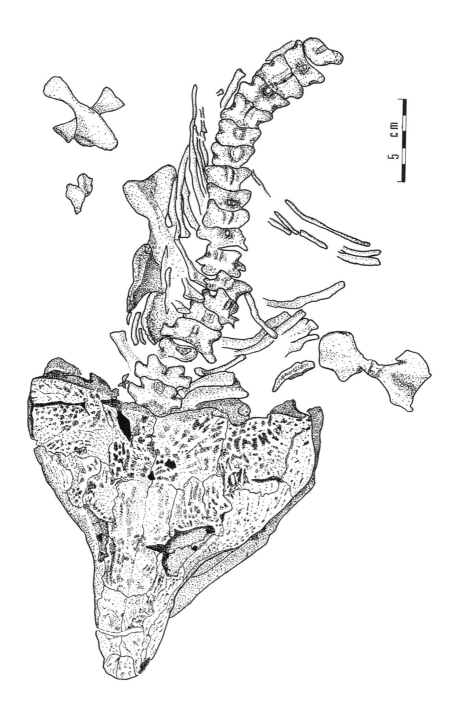

Figure 12. UCLA VP 3200, *Labidosaurus hamatus*, seen mainly in dorsal view.

sentative of all captorhinomorphs. Some specimens of *Labidosaurus* show alternation in neural arch construction and neural spine height; but oddly, other specimens do not.

The number of presacral vertebrae was determined to be 25 by Williston (1917). His observations were based mainly on FMNH UC177, which shows the centra well exposed in ventral view. Re-examination of this specimen and of UCLA VP 3200 confirms that conclusion. Two sacral vertebrae are present and caudal vertebrae appear to number approximately 35 to 40.

Atlas-axis complex. The overall structure of the atlas-axis complex is similar to that in *Captorhinus*, but size differences between the two are substantial. The atlantal centrum is well preserved in UCLA VP 3167: in end view it is oval-shaped, pierced for passage of the notochord, and has a triangular-shaped ventral keel (Plate 2E and F) which is sharp and laterally compressed between its anterior and posterior limits. As in *Captorhinus*, the axial intercentrum is fused to the atlantal centrum, though the ventral keel is more pronounced in *Labidosaurus*. Processes for the axial ribs extend posteriorly from either side of the atlantal centrum. Smooth articular facets for the neural arch pedicels are located on the anterodorsal margins of the centrum. They extend posteriorly to the midlength level of the centrum. Anteroventrally the centrum and its keel are lightly bevelled for reception of the atlantal intercentrum.

The atlantal neural arch in *Labidosaurus* is a paired structure, not fused dorsally (Plate 2C and D). A left atlantal arch is well preserved in UCLA VP 3167, whereas a right atlantal arch is clearly visible in UCLA VP 3200. The neural plate of the atlantal arch is short and curved, with a medially directed articular face at its posterior end that contacted the atlantal centrum, and an anteromedially oriented articular facet that apposed the exoccipitals on each side of the foramen magnum. The base of the neural plate has a ventrally (and slightly posteriorly) directed process for articulation with the tuberculum of the atlantal rib. A constricted neck, measuring about 70% of the length of the neural plate, connects the neural spine to the arch proper. A thin, delicate spine rises above the neck, but did not fuse with its opposite at the midline. Just above and anterior to the neck is a dorsolaterally directed, ellipsoid articulation for the proatlas. Posteriorly the spine tapers off into a narrow wing-like process similar to that in *Petrolacosaurus* (Reisz, 1981). It extends well beyond the posterior zygapophysis of the atlas.

UCLA VP 3167 includes a well-preserved atlantal intercentrum (Plate 2A and B). It is a midventrally keeled crescentic wedge, similar to that in *Captorhinus* and those found in many primitive pelycosaurs (Romer and Price, 1940). Its anterior concave face articulates with the basioccipital, whereas a concave posterior lip contacts the atlantal centrum. A posteriorly angled, buttressed process terminates in the capitular articulation of the atlantal rib.

The proatlas is not preserved in any of the specimens examined, but was surely present, as indicated by the articular facets on the atlantal arch.

UCLA VP 3167 includes a complete, though slightly obscured, axis (Plate 2G), and UCLA VP 3200 includes one that is partially preserved. Both Case (1911) and Williston (1917) stated that the axis differs little from the more posterior presacral vertebrae. However, a much enlarged neural spine, and a laterally compressed centrum terminating ventrally in a sharp keel, belie these statements.

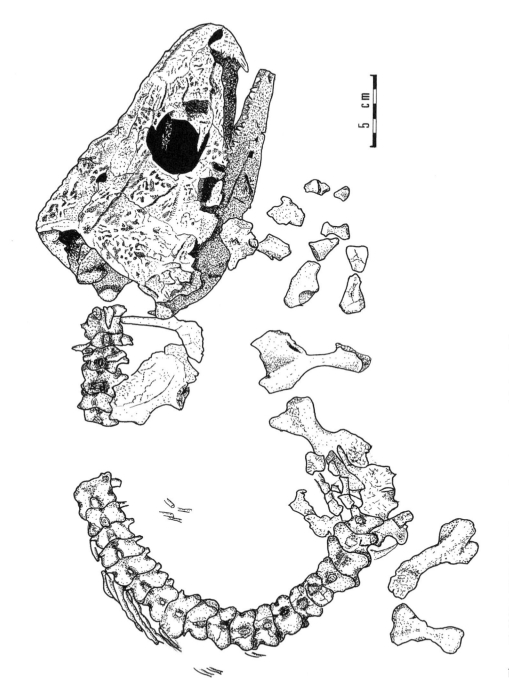

Figure 13. UCLA VP 3167, *Labidosaurus hamatus*, seen mainly in dorsal view.

The axial neural arch is not "swollen" as are those of more posterior presacral vertebrae. The planes of the anterior zygapophyses angle inward at about 30 degrees to the horizontal, whereas the posterior zygapophyses angle ventrolaterally at a 20 degree angle to the horizontal. The transverse processes extend outward from the body of the centrum at angles of 30 degrees to the transverse plane and 10 degrees to the horizontal plane.

The axial neural spine in *Labidosaurus* is impressive in size and proportion, making up as much as one-third of the total vertebral height. Dorsally it is blocky and subtriangular in shape, with the apex directed anteriorly. Its base is constricted, leading to a massive spine that is expanded anteroposteriorly. Scars marking the origin of the M. obliquis capitis magnus and M. rectus capitis posterior, the former smaller than the latter, are particularly prominent. A process at the anterolateral edge of the spine probably served as the origin of the M. spinalis capitis. The posterior face of the neural spine bears two fossae separated by a shallow median furrow demarcated by a sharp ridge at either side. These ridges probably provided attachments for paired Mm. interspinales, and the median furrow may have been the point of attachment of an interspinous ligament.

Dorsal vertebrae. The centra in *Labidosaurus* are spool-shaped, with ventral bevelling of the ends for reception of the intercentra somewhat more pronounced anteriorly than posteriorly. Central width is approximately 10-15% greater than height. The first two dorsals (third and fourth presacrals) possess a ventral keel similar to that of the axis. The average central length increases gradually toward the posterior end of the column. Typically the posteriormost presacral centra are 20-25% longer than the anterior dorsals though there is no substantial increase in their average width. In well-preserved specimens a series of low, parallel, buttressing ridges are visible on the ventral surface of the last five presacral centra.

More anterior transverse processes are well developed, their dorsal edges extending as much as 4.5 mm. beyond the lateral limits of the zygapophyses, giving the tuberculum of the rib a ventromedial orientation of approximately 45 degrees to the horizontal. This angle becomes more vertically oriented posteriorly, and the capitular articulation does not reach as far forward. The last two dorsals bear very small ribs, have correspondingly small articular surfaces, and articulate with the vertebrae via only parapophyses. The pedicels are located more directly over the middle of the centrum than are those in *Captorhinus*.

Many specimens in *Labidosaurus* display a conspicuous and consistent pattern of alternation of neural spine height and neural arch structure, whereas others do not display it at all. *Captorhinus* does not show alternation throughout its column, and therefore it is conceivable that short, isolated strings of vertebrae might not always exhibit the phenomenon. However, a number of complete, articulated specimens of *Labidosaurus* are available and some show alternation throughout the entire presacral column. In other completely articulated specimens however, neural spines are all of approximately equivalent height. Thus, there are essentially two different vertebral "morphs" in *Labidosaurus*, which can be referred to as the "alternating morph" and "non-alternating morph"; both morphs are currently assigned to the same species, *L. hamatus* (Sumida, 1987). Appendicular elements associated with the two different morphs are virtually identical (Sumida, 1989a), and potential specific distinctions await a detailed examination of cranial materials.

In both morphs the neural arches are conspicuously "expanded" or "swollen," rising up on either side from the level of the posterior zygapophyses to meet medially at the base of the neural spine. Distinct, anteriorly facing pits suggest that very well-developed Mm. interarticulares attached at the lower lateral aspect of the arch and posterior zygapophysis. Just dorsal and medial to the pits on each side of the neural arch is a shallow anterolaterally facing depression that follows the curvature of the arch. These depressions are more distinct in the anterior half of the column. By comparison with *Captorhinus*, they were apparently points of attachment for the M. spinalis dorsi and M. semispinalis dorsi. The neural arches of the last two dorsals show posteriorly directed, nipple-shaped processes, possibly points of insertion for the M. longissimus dorsi. The zygapophyseal planes are slightly angled, tilting from 5 to 10 degrees.

The neural spines of the non-alternating morph (Plate 2H) are much like those described by Case (1911) and Williston (1917). Case stated: "The neural arches are low, wide and swollen so that the upper surface looks almost hemispherical. The spines are low and short, bifurcate in anterior vertebrae . . ." Williston added: "The spines anteriorly are a little more slender, giving greater freedom of vertical movement." This picture of the non-alternating morph has been widely accepted as describing the vertebral structure in *Labidosaurus* (e.g., UCLA VP 3200, FMNH UC726, AMNH 4417).

UCLA VP 3167, FMNH UC178, FMNH P12758, USNM 17045, MCZ 8923, and other partial columns display the alternating morph (Plate 2H, I, and J). Tall, wide, conically shaped spines alternate with low, narrow, wedge-shaped spines. The low type extends through the anteroposterior length of the neural arch, whereas the tall type slopes upward steeply, ranging from 50% to 300% higher than adjacent low types. Tall-type spines typically constitute about 20% of the total vertebral height, while the low spines never constitute more than 10%. Alternation is superimposed on a slightly decreasing average spine height posteriorly along the column, but marked differences in spine height reoccur just anterior to the pelvic girdle.

Tall-type neural spines are significantly wider, almost cylindrical at their bases; low types are often less than one-third the width of an adjacent tall type. Tall-type neural spines may be distinguished even when they are broken off, as they leave a circular break mark, in contrast to the long, narrow, rectangular break left by the low type.

As neural spines are often broken off, description of their internal structure is possible. Tall types show a core of spongy bone bordered by compact bone which is characteristically thicker anteriorly and posteriorly where Mm. interspinales were attached. Trabeculae are frequently preserved in tall-type spines, but few or no trabeculae can be seen in the narrow low-type spines. A number of the tall-type spines are bifid. This bifidness appears to be almost random, but occurs a bit more frequently anterior to the limb girdles.

Neural arches of the low type have a distinct furrow just lateral to the neural spine, creating an anteroposteriorly directed trough (Figure 14C). The portion of the arch just lateral to this trough is more rounded than in the tall type, where the arch rises steeply to the higher spine. It seems likely that paired Mm. interspinales passed between the tall-type spines through these furrows. The zygapophyseal planes of the alternating

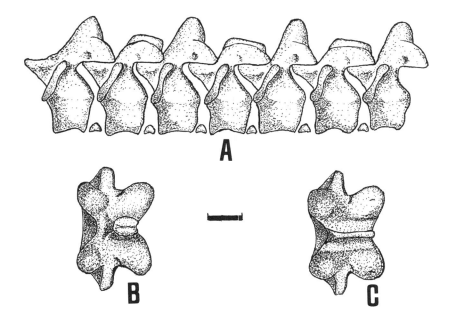

Figure 14. Reconstruction of presacral vertebrae of *Labidosaurus hamatus*; illustrations are somewhat idealized. A, reconstruction of dorsal vertebrae of the alternating morph of *Labidosaurus hamatus*, left lateral aspect of articulated vertebrae and intercentra. Based on MCZ 8923 and UCLA VP 3167. B, single dorsal vertebra, tall type, dorsal aspect, anterior to the left. Based on middorsal vertebrae of MCZ 8923 and FMNH UR161. C, single dorsal vertebra, low type, dorsal aspect, anterior to the left. Based on middorsal vertebrae of MCZ 8923 and UCLA VP 436.

morph are more tilted than those of the nonalternating form, with angles ranging from 10 to 15 degrees.

A consistent sequence has been observed in those specimens that exhibit alternation. The first dorsal (third presacral) vertebra is of the tall type, and alternation follows back to the last presacral. The pattern is not as perfect in *Labidosaurus*, however. There are two interruptions in the pattern, presacrals 15 and 16; and 21 and 22 are of the tall type.

Sacral vertebrae and ribs. Sacral structures are well preserved in FMNH UC726 (Plate 2I and J), UCLA VP 3200, and USNM 17045. As in *Captorhinus*, the first sacral vertebra can be easily distinguished by the articulation of the massive sacral rib and narrowing of the neural arch. Although the anterior zygapophyses are as wide as those of the posterior presacrals, the posterior zygapophyses of the first sacral vertebra measure only 65-70% of that width. The neural arch of the second sacral is similarly narrow.

The neural spine of the first sacral is as tall as that of the last presacral, the second slightly shorter. Though tall, they are laterally compressed; they show no forward tilt as in *Captorhinus*. Neither neural arch shows furrows lateral to the neural spine.

The sacral centra have a particularly heavy build; the first is approximately 25% longer than the most posterior dorsals, and the second a bit shorter than the first. The centra of both are bevelled for reception of the intercentrum, which is wedged in rather tightly, but not fused to the centrum. Like that of *Captorhinus*, the articulation between the first sacral and the last presacral is extremely firm from the midcentrum dorsally (Plate 2I and J). This is partially accomplished by the anterior expansion of the neural arch pedicel so as to allow for an extremely large costal articulation. The near ankylosis between the first sacral and last dorsal vertebrae appears to indicate that sacral movement was tightly coupled to that of the most posterior dorsal vertebrae. Ventrally neither sacral centrum shows the pattern of buttressing ridges found in the posterior presacrals.

The articular facet for the first sacral rib has an almost trapezoidal shape, with a prominent apex directed more ventrally than that in *Captorhinus*. In mature specimens the area of articulation approaches 1.2 sq. cm. The head of the rib shows little constriction into a neck proximal to its distal expansion into a broad, spatulate structure that articulates with the ilium. The second sacral rib is not nearly as massive, measuring only one-third to one-half the diameter of the first proximally. It is directed laterally and slightly anteriorly to its articulation with the ilium and a posterior embayment of the first sacral rib.

Caudal vertebrae and ribs. The number of caudal vertebrae is, as usual, uncertain. Broili (1908) reported only 17, Williston (1917) restored *Labidosaurus* with 25, and Romer (1956) estimated that most captorhinids had approximately 40. Fox and Bowman (1966) agreed with Romer's estimate for *Labidosaurus*, but Heaton and Reisz (1980) consider 60 to be a more realistic estimate of the number of caudal vertebrae. USNM 17045 and UCLA VP 3491 (Plate 2) bear directly on the question of the number of caudals: the former is nearly complete, and 33 can be counted; the latter has 35. In the specimens examined, the tail tapers down to a small tip by the 35th caudal, so it seems that an estimate of 60 might be too high for *Labidosaurus*. A conservative estimate would be 35 to 40 caudals. The anteriormost caudals closely resemble the second sacral vertebra in having narrow neural arches and spines that become more attenuated posteriorly. The zygapophyses are not as strongly buttressed as those of the dorsal or sacral series, and their articular planes tilt inward at an angle of 15-25 degrees to the horizontal.

The first 4 caudals bear short, posteriorly recurved ribs. Subsequent vertebrae have posteroventrally directed haemal spines, the bases of which are accommodated by a bevelling of the ends of the centra. Beyond the anteriormost few, the caudal centra become progressively more slender and elongate, whereas the haemal spines become more markedly shorter.

Intercentra. Small, wedge-shaped intercentra are found throughout the dorsal column except between the atlantal and axial vertebrae. As preserved, they do not fill the intercentral gaps completely, and were probably completed by cartilage. The more posterior intercentra are somewhat larger.

Presacral ribs. Case (1911) stated that the last 4 or 5 dorsals do not bear ribs, but ribs can be observed on all of the presacral vertebrae as well as the sacrals and first 4 caudals. In all but possibly the axial rib the tuberculum and capitulum are not clearly separated, but joined by a thin web of bone. The first 2 ribs are short and thin, and their distal ends are only slightly expanded. The 3rd is much heavier, and the 4th heavier still, with its shaft enlarged into a broad spatulate structure. The 5th is larger than the following ribs, but less so than the 4th. As in *Captorhinus* (Holmes, 1977), these modified ribs undoubtedly functioned in support of the pectoral girdle via a well-developed serratus musculature. Posterior to the 5th rib, the shafts decrease in diameter. In the mid-dorsal region they become progressively shorter and begin to cant posteriorly. Mid-dorsal ribs are about three-fourths the length of the 5th and 6th ribs.

The ribs of the last few vertebrae are directed posteriorly at a sharp angle and are quite small, only about one-fourth the length of the longest ribs. The tubercular attachment is absent, leaving only a small, rounded capitulum. This condition is particularly evident on FMNH UC 726 (Plate 2I and J). The diapophysis of the 25th vertebra is larger than the two anterior to it, and faces posteriorly toward the pelvic girdle. The last dorsal rib may have aided in sacral support via a ligamentous connection, also helping to couple sacral movement to that of the posterior dorsal vertebrae.

PROTOCAPTORHINUS

Case (1902) based his original description of *Pleuristion* on a short string of vertebrae that were characterized by "broad winglike transverse processes, and by the unusually broad, large neural canal," but Olson (1970) noted that these characters alone are not sufficient to distinguish *Protocaptorhinus* (=*Pleuristion*) from other captorhinids. With the declaration of *Pleuristion brachycoles* (Case, 1902) as a *nomen dubium* and subsequent assignment of its materials to *Protocaptorhinus pricei* (Clark and Carroll, 1973) by Olson (1984), *Protocaptorhinus* is known from a wide spatial and stratigraphic range, including the Upper Permian of Zimbabwe in southern Africa (Gaffney and McKenna, 1979), and possibly India (Kutty, 1972). However, it is the type specimen and abundance of material from Lower Permian strata near Orlando, Oklahoma (Olson, 1984), that now permits a more detailed description of the vertebral column.

Atlas-Axis Complex. Elements of the atlas and axis are known only from the type specimen, MCZ 1498. The general structure of the components is similar to that of *Captorhinus*, though not as heavily constructed. Only a small portion of the proatlas is accessible. What can be seen is thin and wedge-like. The atlantal intercentrum is a crescent-shaped wedge with no visible indication of an articular process for the capitulum of the atlantal rib. However, distinctly bicipital ribs in this region make an assumption of its presence in life reasonable. Only the dorsal plate of the right atlantal neural arch is preserved, and it is essentially similar to that in *Captorhinus*. Contrary to the reconstruction by Clark and Carroll (1973), the presence of a posteriorly directed spine of the atlantal arch is indicated by a caudally directed process.

The atlantal centrum is lozenge-shaped, with a blunt, posteriorly directed parapophysis, presumably derived from a fused axial intercentrum as in *Captorhinus*.

The axial centrum is similar in length to the centra of the following vertebrae and has a large, laterally directed diapophysis and a rounded ventral keel.

The axial neural spine in *Protocaptorhinus* overhangs the neural arch of the atlas anteriorly. It is blocky in lateral view and angles anterodorsally, in contrast to the anteroventral angulation usually seen in *Captorhinus*. However, as only one example of the atlas-axis complex is known for *Protocaptorhinus*, the potential variability of axial structure is not known.

Dorsal vertebrae. Assuming that materials assignable to this genus show serial changes in the structure of the transverse processes similar to those in *Captorhinus* and *Labidosaurus*, placement of isolated individual (or short strings of) vertebrae in the column can be made. This in turn permits description of regional differences in the vertebral column.

In end view, the centra of *Protocaptorhinus* (Figure 15) are more rounded than those in *Captorhinus* or *Eocaptorhinus*. In the presence of ventral keels as far posteriorly as at least the 5th or 6th presacral, *Protocaptorhinus* is similar to *Eocaptorhinus*. The ventral keel may have served as a strengthening brace for the centrum in more primitive captorhinids. Centra in the posterior two-thirds of the column are bevelled for reception of intercentra.

The transverse processes are essentially like those of *Eocaptorhinus* and *Captorhinus*. In most of the specimens examined the neural canal is large relative to the height and width of the centrum: The width of the canal approaches 85-90% that of the centrum. The outline of the neural canal in the anterior dorsal vertebrae is kidney-shaped. In the more posterior dorsal vertebrae of *Protocaptorhinus*, the outline of the neural canal becomes almost trapezoidal, with the apex directed ventrally. The lateral walls of the neural arch in *Protocaptorhinus* are fairly thin, and the arch is not as conspicuously swollen throughout the length of the column as are those in *Eocaptorhinus* or *Captorhinus*.

In the anterior portions of the column the neural arches are situated markedly posterior relative to the centrum (Olson, 1970: Fig. 3F). The pedicels become stouter and more elongate anteroposteriorly. The neural arches in *Protocaptorhinus* (Figure 15) are less swollen and longer anteroposteriorly than those in *Captorhinus* and *Eocaptorhinus*, but more heavily built than those in *Romeria*. This gradual increase in the stoutness of the neural arch parallels Heaton's (1979) scheme of the phylogenetic development of the captorhinid skull. There is no evidence of mammillary processes of the neural arches. The articular planes of the anterior and posterior zygapophyses in *Protocaptorhinus* have a slight tilt. The anterior pair are mildly cup-shaped, receiving the complementary structures of the posterior zygapophyses.

Clark and Carroll (1973) did not detect alternation of neural spine height in their initial description of the anterior portion of the column in *Protocaptorhinus*. However, UCLA VP materials that display alternation are from more posterior portions of the column, a condition similar to that reported for *Eocaptorhinus* (Heaton and Reisz, 1980).

The neural spines in *Protocaptorhinus* (Figure 15) are more slender than the typical cone-shaped structure of *Captorhinus* tall-type vertebrae. Those tall spines with dorsal tips that have broken away display a thick, well-developed posterolateral wall,

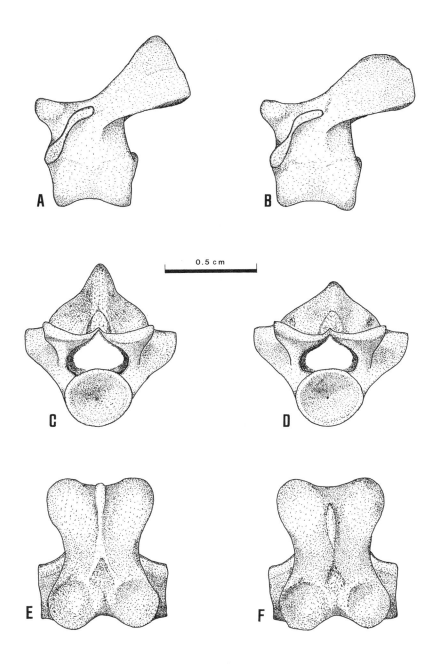

Figure 15. Dorsal vertebrae of *Protocaptorhinus pricei*. A, C, and E, tall-spined vertebra. B, D, and F, low-spined vertebra. A and B are left lateral aspects, C and D anterior aspects, E and F dorsal aspects. A and B based on UCLA VP 3531 and UCLA VP 3539. C based on UCLA VP 3531, UCLA VP 3539, and UCLA VP 3621. D based on UCLA VP 3539, UCLA VP 3621, and UCLA VP 3685. E and F based on UCLA VP 3537, UCLA VP 3539, and UCLA VP 3627.

much like the condition found in *Captorhinus*. The low-spined vertebrae of *Protocaptorhinus* are not as sharply wedged as those of *Captorhinus*. Nonetheless, those with broken-off tips do not show any sign of trabeculae internally. The absolute and relative differences in height between high- and low-type spines of *Protocaptorhinus* are not as great as in *Captorhinus*.

The points of attachment for the M. interspinales, M. spinalis dorsi, M. semispinalis dorsi, and Mm. interarticulares are similar to those in *Captorhinus*. No other muscular attachments can be determined with certainty for *Protocaptorhinus*.

It is impossible to reconstruct a column-long pattern of alternation for *Protocaptorhinus*, as no complete presacral columns exist. Alternation of neural spine height is probably more common in the posterior portions of the column.

Sacral vertebra and rib. Olson (1970) indicated that *Protocaptorhinus* possessed only one sacral vertebra. All other captorhinids for which complete sacral structures have been described have two sacral vertebrae. Certain protorothyridid genera are known to have one or two sacral vertebrae, depending on the genus (Clark and Carroll, 1973; Carroll and Baird, 1972); thus, different sacral counts within the Captorhinidae would not be surprising. UCLA VP 3533 (probably an immature individual) includes one extremely worn sacral vertebra that is not significantly different from the first sacral of other captorhinids. The neural spine, though well developed, shows no indication of contributing to any pattern of alternation of neural spine height.

The single sacral rib in *Protocaptorhinus* is well developed, with little constriction between its base and distal contact with the ilium. The rib immediately caudal to the definitive sacral rib is slender and recurved, but shows a significant lateral extension. The rib immediately anterior to the sacral rib is also quite slender. It is not recurved, but is directed posterolaterally toward the general region of the ilium and may have helped to support the pelvic girdle via a ligamentous connection.

Caudal vertebrae and ribs. The scattered nature of the available materials prohibits a determination of the actual number of caudal vertebrae. Well-developed caudal centra may be seen in UCLA VP 3546 (Figure 16) and are much like those in *Captorhinus*. Though lighter in build, caudal centra are essentially cylindrical in outline like their dorsal counterparts. The first few caudal ribs are sharply recurved.

Presacral ribs. The first three ribs in *Protocaptorhinus* are slender, and only the third gives any impression of a thickening of the shaft (Clark and Carroll, 1973). Its distal end is spatulate, but not to the degree of the comparable ribs in *Captorhinus* or *Labidosaurus*. Complete ribs are almost never preserved in the nodules recovered from the Orlando locality. In all of the specimens that can be observed the rib heads are holocephalous, but the capitular and tubercular expansions are well developed, connected by only a very thin web of bone (Figure 16C).

Intercentra. Where preserved, the intercentra are blunt wedges, very similar in construction to those in *Captorhinus*. They are found in the posterior two-thirds of the presacral column and with the anterior caudals. A couple of well-preserved caudal chevrons can be seen in UCLA VP 3596 (Figure 16). Their vertebral articulations are curved, face dorsomedially, and are slightly cup-shaped.

Although *Protocaptorhinus* exhibits similarities to *Eocaptorhinus, Captorhinus,* and certain protorothyridids in the construction of its vertebrae and ribs, it does not

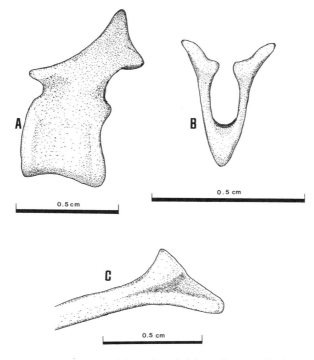

Figure 16. *Protocaptorhinus pricei.* A, caudal vertebra, left lateral aspect. B, chevron, anterior aspect. C, left presacral rib head and proximal portion of the shaft; posterior aspect. A based on UCLA VP 756, UCLA VP 3531, and UCLA VP 3621. B based on UCLA VP 3546. C based on UCLA VP 3621.

appear to be easily referrable to any of them. Although Heaton (1979) noted that *Protocaptorhinus* might be appropriately placed in synonymy with either *Romeria* or *Eocaptorhinus,* the description presented here agrees with Heaton's placement of *Protocaptorhinus* in a position intermediate between these two genera. The structure of the ribs and vertebrae seems to reinforce the conclusions of Clark and Carroll (1973) and Olson (1984) that it constitutes a distinct genus.

CAPTORHINIKOS

The genus *Captorhinikos* was first described by Olson (1954) on the basis of cranial materials and a small amount of postcranial material. He distinguished two species, *Captorhinikos valensis* and *C. chozaensis*, on the basis of dentition, lower jaw structure, and a difference in development of the neural spines. Olson (1970) later described a third and smaller *C. parvus* from the Hennessey Formation in Oklahoma.

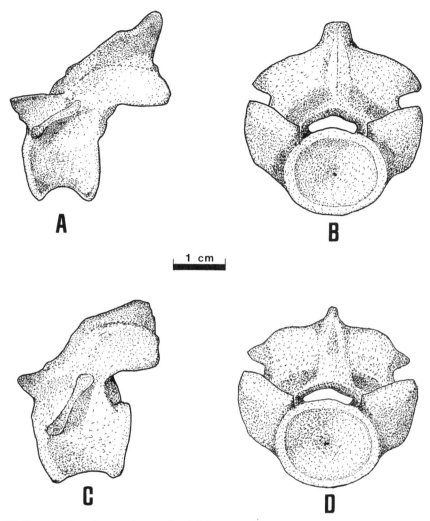

Figure 17. *Captorhinikos chozaensis*. A and B, left lateral and anterior views of tall-spined vertebra. C and D, left lateral and anterior views of low-spined vertebra. A and B based on FMNH UR15; C and D based on FMNH UR225.

Captorhinikos chozaensis

No specimens of *Captorhinikos chozaensis* are complete enough to permit a thorough description of the postcranial skeleton, but USNM 21275, FMNH UR857, and FMNH UR858 allow description of most of the components of the vertebral column except for the atlas-axis complex. Olson (1970) estimated the presacral count at about 25, but also noted that it may have had more.

Dorsal vertebrae. The dorsal vertebrae of *C. chozaensis* (Figure 17) have been described as similar to those of *Captorhinus* except for a proportionately better-developed neural spine in the former (Olson, 1954, and in Olson and Barghusen, 1962; Vaughn, 1958). On the average, the vertebrae of *C. chozaensis* are about twice as large as those of mature specimens of *Captorhinus*, and better-developed neural spines may be a consequence of its larger size.

Centra, pedicels, and zygapophyses of *C. chozaensis* have a typically captorhinid construction. The zygapophyses are widely spaced, somewhat cup-shaped, and slightly tilted. The lateral extents of the anterior zygapophyses are marked by a well-rounded lip, and their concavities extend to the base of the neural arch.

The neural arches of *C. chozaensis* do not extend as far back beyond the posterior limit of the centrum as do those in *Captorhinus* or *Protocaptorhinus*. Neural spines are well developed in most cases, and there is a clearly visible alternation in neural spine height and arch structure in certain specimens. Low-type neural spines of *C. chozaensis* are considerably narrower and shorter than adjacent tall types, often less than half their height. They are somewhat nubbin-shaped.

The neural arches of vertebrae with taller neural spines do not appear as widely swollen as those of low-spined vertebrae. Tall-type spines are well developed. Anteriorly they are not quite as robust as those more posterior, which are rounded and knobby in construction.

USNM 21275 indicates that alternation occurred in the anterior portion of the vertebral column, probably near or above the pectoral girdle. UCLA VP 3805 is a low-spined vertebra from the middle of the presacral series. A definite pattern of alternation cannot be established with available materials, but the evidence suggests that alternation probably occurred in most parts of the presacral column.

Sacral vertebrae and ribs. As the sacral vertebrae and ribs of *C. chozaensis* have been described and figured by Olson (1970), little needs to be added here other than that they display a typical captorhinid pattern.

Caudal vertebrae and ribs. FMNH UR857 includes 15 caudal vertebrae, but the tail was surely much longer. The neural arches are not swollen. Caudal neural spines, though well developed in the proximal tail region, are slender, showing no evidence of alternation of height or structure. The caudal series is similar to that of *Labidosaurus* in other features.

Presacral ribs. As in other captorhinids, the ribs of *C. chozaensis* are holocephalous, but unlike other members of the family there is no spatulate expansion of the shafts near the pectoral girdle. Mid-dorsal and posterior ribs are more slender than those anterior, and remain essentially circular in outline. The posteriormost vertebrae in USNM 21275 do not show distinct costal articulations, but the specimen is badly weathered, making determination of the presence of ribs difficult.

Captorhinikos parvus

Captorhinikos parvus was originally described on the basis of abundant materials from the Hennessey Formation of Cleveland County, Oklahoma (Olson, 1970). Enough postcranial material was present to allow Olson to conclude that the high degree of ossification in the smaller *C. parvus* warranted its specific distinction. Although

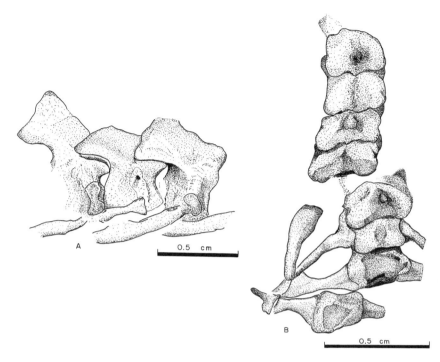

Figure 18. *Captorhinikos parvus*. A, UCLA VP 2985; right lateral view of dorsal vertebrae. B, UCLA VP 2896; dorsal view of last 5 presacral, 2 sacral and 1st caudal vertebrae. Anterior toward top of page.

Ricqles (1984) has questioned the validity of *C. parvus*, it is retained here as a taxonomic entity until the genus is carefully restudied as a unit.

None of the available materials contains examples of the atlas-axis complex or most of the caudal vertebrae, but UCLA VP specimens include substantial portions of presacral column that reveal heretofore undescribed details of its structure. No confident estimates of the actual number of presacral or caudal vertebrae can be made at this time.

Dorsal vertebrae. Captorhinikos parvus is among the smallest of the captorhinids and its vertebrae are correspondingly proportioned. The centra of *C. parvus* are notochordal and deeply amphicoelous. Fortuitously fractured centra show the notochordal foramen to be quite small. There is no indication of a ventral keel.

The pedicels of more anterior vertebrae are distinguished by anterior and posterior constrictions above the centrum, which more precisely delimits the anterior zygapophyses and neural arches. More posteriorly there is little or no narrowing of the pedicel relative to the centrum. In fact, the anterior zygapophyses of more posterior vertebrae are almost completely supported ventrally by the pedicel, a condition not seen in other captorhinids. The zygapophyses are just slightly cup-shaped.

The neural arch is placed posteriorly on the centrum so that the neural spine extends over the anterior half of the following centrum (Figure 18A). As in other captorhinids, the neural arches of the anteriormost presacrals are not as heavily swollen as those farther back. None of the neural arches of *C. parvus* are as conspicuously swollen as in most other captorhinids, possibly a reflection of its small size. Among the captorhinids, only *Romeria* possesses narrower neural arches. *Captorhinikos parvus* did not have the proportionately large, heart-shaped skull of other advanced captorhinids, and the need for the support that may have been afforded by heavier, more swollen neural arches may not have been as great.

Alternation of neural spine height is distinct in the posterior vertebral column of *C. parvus*. If we assume that it had a typical captorhinid presacral count of 25, presacrals 24 and 22 would have been low-spined types, a pattern similar to that in *Captorhinus*. Low spines in this portion of the column of *C. parvus* have the shape of narrow wedges that are barely visible in lateral view (Figure 18A) and except for a fine line marking their midline they are even less visible dorsally (Figure 18B). Adjacent taller neural spines are somewhat compressed laterally and restricted to the posterior of the neural arch. They are not as conical as those of *Captorhinus, Eocaptorhinus,* or *Labidosaurus*, but nevertheless remain structurally distinct from those of adjacent low types.

An occasional low-spined vertebra is found in the mid-column region, but it is not as distinctly different from adjacent vertebrae as are those of the more posterior portion of the column. The difference is more in lateral width than height (as in some specimens of *Petrolacosaurus*, see Reisz, 1981). Neural spines of the mid-dorsal region are well developed, triangular in lateral view, and terminate in a well defined apex. Though relatively taller and more strongly developed than those of the posterior region, they are also narrower in lateral measure. The neural spines and arches of the anterior vertebrae are not preserved well enough to permit description of possible patterns of alternation in that region.

Sacral vertebrae and ribs. UCLA VP 2896 includes remains of the sacral structures of *C. parvus*. It had two sacral vertebrae with associated ribs. They are similar to those of *Captorhinus* with only minor proportional distinctions, although the second sacral rib is stouter and better developed than those of other captorhinids.

Caudal vertebrae and ribs. Almost no material from the tail of *C. parvus* is available for study, but UCLA VP 2895 does include the first caudal and portions of its associated ribs. The vertebra is chipped and quite worn, but does indicate proportions similar to those of the second sacral and anterior caudals of other small captorhinids. The first caudal rib is unusually heavily built, not strongly recurved, and projects directly toward the ilium. The first caudal rib might have lent accessory sacral support (Figure 18B), but because the specimen is not complete this cannot be determined with certainty.

Presacral ribs. Because of the nature of preservation, most of the ribs in the UCLA VP specimens are distorted, and none are complete. No significant departures from the basic captorhinid pattern, other than those described for the sacral and caudal structures, can be seen.

Captorhinikos valensis

Captorhinikos valensis was distinguished from the more advanced *C. chozaensis* on the basis of the jaw's being more slender for the length of its tooth-bearing portion (Olson, 1954). As with the type specimen of *C. valensis*, referred presacral elements are fragmentary and extremely worn. Available specimens do not include components of the atlas-axis complex or sacral and caudal portions of the column.

The worn condition of FMNH UR106 and FMNH UR108 does not allow detailed description of surface structure or muscle scars, but these specimens do indicate that the vertebrae of *C. valensis* are proportionally similar to those of *C. chozaensis*. The neural arches of *C. valensis* are broad, though not as rounded as those of *Captorhinus, Eocaptorhinus,* or *Labidosaurus*. They resemble the neural arches of *Protocaptorhinus*, but are more laterally expanded.

Olson (1954) emphasized well-developed neural spines as one of the few distinctive postcranial features in the genus *Captorhinikos*. The neural spines of *C. valensis* are intermediate in development between those of *Captorhinus* and *Captorhinikos chozaensis*. FMNH UR106 includes 7 approximately mid-dorsal vertebrae. The 1st neural spine of this series is completely missing, but the next 6 show a perfectly regular pattern of alternation. Unfortunately, they are quite worn, and the presence or absence of furrows lateral to the low-type neural spines cannot be determined. FMNH UR108 is likewise worn. Only the first 3 vertebrae of its string of 6(?) are preserved well enough to allow even a cautious attempt at determination of neural spine structure. The first 3 neural spines appear to alternate in a high-low-high pattern. The available specimens do little to establish an overall pattern of alternation for *C. valensis*, but they do confirm its presence.

Captorhinikos: Summary

Alternation of neural spine height is present in all three species assigned to *Captorhinikos*. Despite the inability to determine the exact patterns of the phenomenon, it was surely present in at least the posterior (just anterior to the pelvic girdle) and mid-dorsal regions of the column in the genus *Captorhinikos*.

LABIDOSAURIKOS

Vertebral remains of the genus Labidosaurikos are extremely rare. Their generic assignment is often tenuous, and specific assignment even more so. The type species, *Labidosaurikos meachami* (Stovall, 1950) included no postcranial elements. Olson (1954) described *L. barkeri* on the basis of fewer maxillary and dentary tooth rows. The type specimen of *L. barkeri*, FMNH UR110, does include 10 presacral vertebrae. Seltin (1959) synonymized the two species, considering *L. barkeri* an immature growth stage of *L. meachami*. Olson and Mead (1982) retained Olson's (1954) earlier assignment, and their conclusions are tentatively accepted here in lieu of a more complete study of the genus. Except for FMNH UR110, all other vertebral elements of *Labidosaurikos* may be assigned with confidence at only the generic level.

Only elements of the mid-dorsal and posterior dorsal regions are known for *Labidosaurikos*. In addition to the type of *L. barkeri*, UCLA VP 577, FMNH UR373,

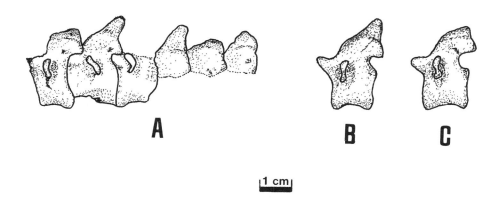

Figure 19. UCLA VP 577; *Labidosaurikos* sp. A, left lateral view of dorsal vertebrae. B and C, reconstruction of tall- and low-spined vertebrae.

FMNH UR15, and FMNH UR225 also include vertebral elements. The dorsal vertebrae of *Labidosaurikos* (Figure 19A-C) are similar in size and shape to those in *Labidosaurus*. No specimens are complete enough to allow an estimate of the presacral count.

In end view the centra in *Labidosaurikos* are slightly ovoid and the edges are smoothly rounded, with ventral beveling for the intercentra. There is no distinctly defined keel. The transverse processes of mid-dorsal vertebrae are of the characteristic captorhinid pattern. *Labidosaurikos* probably possessed ribs throughout its presacral series, but an absolute determination cannot be made due to the weathered condition of the available specimens.

The articular planes of the zygapophyses show a small, but definite tilt. The articular surfaces are flat, but not horizontal. The neural canal appears more dorsoventrally compressed than in most other captorhinids, but it is difficult to determine this exactly.

Alternation of neural spine height and neural arch structure can be seen clearly in all of the *Labidosaurikos* specimens examined. Low-type neural spines are narrow, wedge-shaped, and approximately 40% the height of adjacent tall-type spines. Low-type spines extend the entire anteroposterior length of the neural arch, with the posterior half having a dorsally projecting ellipsoid tip. They bear clearly defined furrows lateral to the spine in a manner similar to that in *Labidosaurus*. Tall-type spines are subcircular in frontal section, terminating in a knobby tip, and more than 50% wider than low types. No lateral furrows can be seen in the neural arch of the taller spines. Low-type spines appear slightly triangular in lateral view. Taller spines are also subtriangular in lateral view, though less acute. Low-type spines have virtually no area for attachment of Mm. interspinales. Tall-type spines have substantial areas of attachment for Mm. interspinales, M. spinalis dorsi, and M. semispinalis dorsi. Well-developed pits, presumably for the attachment of Mm. interarticulares, are situated on the anterolateral aspect of the neural arch near the posterior zygapophyses.

Intercentra. Intercentra may be seen in UCLA VP 577 and presumably were present throughout the presacral series. They are somewhat crescentic in shape. All of their edges are rounded.

Ribs. UCLA VP 577 preserves just the heads of a couple of ribs. They do not differ significantly from those of other captorhinids.

THE FAMILY CAPTORHINIDAE: SUMMARY

Most species of Paleozoic reptiles known to display alternation of neural spine height and neural arch construction are members of the family Captorhinidae. Other groups, to be described below, also exhibit this phenomenon, but captorhinids provide the clearest basis for comparison. Their vertebrae are easily recognized, but the concept that they are homogeneous must be reconsidered. Certain distinct trends can be seen:

1. Larger forms tend to have better-developed neural arches and neural spines.

2. Primitive forms have relatively narrower neural arches. Forms such as *Protocaptorhinus* and *Romeria* have neural arches that are actually more extended anteroposteriorly. Those of more derived forms (in the analysis of Gaffney and Mckenna, 1979 and Ricqles, 1984) tend to be much wider, their lateral width being much greater than their anteroposterior length.

3. The development of heavily built neural arches seems to parallel the increase in size of the large, heart-shaped head of captorhinids. This increase seems to reinforce the phylogenetic pattern of changes in skull dimensions outlined by Heaton (1979), but it may also have to do with the increase in overall body size of more advanced captorhinids.

4. Alternation of neural spine height and neural arch structure is evident in almost every member of the family. Its apparent absence in some taxa simply may be due to the paucity of available materials.

5. Alternation of neural spine height consistently occurs near at least one of the limb girdles. The most primitive expression of alternation seems to occur near the pelvic girdle (e.g., in *Protocaptorhinus* and some specimens of *Eocaptorhinus*). In more derived forms, the phenomenon extends into the anterior region of the column, adjacent to the pectoral girdle. In forms where the complete pattern of alternation is not known, it still seems to occur consistently near the limb girdles (e.g., in *Captorhinikos*).

6. In regions of the vertebral column where alternation of neural spine height is distinct, zygapophyseal articular facets are slightly tilted and probably allowed more columnar dorsiflexion than has been considered possible previously.

7. In taxa where sacral structure is available for study, the sacral vertebrae are tightly linked to the last presacral vertebra, requiring the sacrum to respond to the movements of the posterior presacral vertebrae.

8. Examination of muscle scars and histological sections indicates that in areas of alternating high and low spines a system of paired Mm. interspinales passed between tall-type neural spines through furrows on either side of low-type neural spines. Trabecular structure supports this, as well as an alternating attachment of M. spinalis dorsi and M. semispinalis dorsi to only tall-spined vertebrae in regions of alternation.

3

SEYMOURIAMORPHA AND DIADECTOMORPHA

The Seymouriamorpha and Diadectomorpha have been variously classified, but most phylogenetic analyses have consistently pointed out their close relationship, even though other forms grouped with them have been quite diverse (Case, 1911; Watson, 1917a; Olson, 1947). Romer (1964) noted that *Diadectes* should probably be regarded as an amphibian, a conclusion that also pointed out the amphibian nature of the more primitive *Seymouria*. Vaughn (1964) and Moss (1972) described *Tseajaia* as an important transitional form between *Seymouria* and *Diadectes*. In his redefinition of the Cotylosauria, Heaton (1980) re-emphasized the close relationship between the seymouriamorphs and diadectomorphs, including the suborder Seymouriamorpha (composed of the Nycteroleteridae, Kotlassiidae, and Seymouriidae) and the Diadectomorpha (composed of the Limnoscelidae, Tseajaiidae and Diadectidae). Considering the checkered history of the term "Cotylosauria," the classificatory scheme proposed by Panchen (in Panchen and Smithson, 1988), and reiterated in part by Gauthier et al. (1988), is preferred here. The latter authors defined the Seymouriamorpha and Diadectomorpha as basal members of the Batrachosauria. These interpretations are reflected eclectically in Figure 2, and the close relationship of these groups is emphasized here. Accordingly, their vertebral structure is considered together.

Members of the Nycteroleteridae and Kotlassiidae have not been reported to display alternation of neural spine height and structure (Bystrow, 1944; Efremov, 1946), but some form of the phenomenon is expressed in certain taxa of all the other groups listed above. Members of these groups are among the most fully terrestrial of Paleozoic amphibians (Carroll, 1988). They have played an important role in studies of the amphibian-to-reptile transition and in phylogenetic analyses of both groups, commonly serving as an outgroup for phylogenetic studies of amniotes (Heaton, 1980; Brinkman and Eberth, 1983; Heaton and Reisz, 1986). They are the most primitive groups to show alternation of neural spine structure (possibly excepting certain microsaurs).

Heaton (1980) noted that the vertebrae of seymouriamorphs and diadectomorphs are quite similar and that the far lateral extent of the neural arch (i.e., the "swollen" or "expanded" neural arch) may be a primitive character of both "cotylosaurs" and early reptiles (see also Heaton and Reisz, 1986). It must be noted that *Seymouria, Tseajaia*, and *Diadectes* seem not to be direct ancestors of reptiles (Olson, 1947, 1965). The fact that they were fully terrestrial animals, despite their classification as amphibians (Olson, 1976), is more significant for this study.

SEYMOURIAMORPHA

The seymouriamorphs include a number of European and Russian forms, but only *Kotlassia* (Bystrow, 1944) is well known from the literature. It is not known to display alternation of neural spine height, and the European forms were not available for study. The focus of this portion of the study centers on the well-known North American genus *Seymouria*.

Seymouriidae: *Seymouria*

The genus *Seymouria* is best known from its most common representative, *S. baylorensis*. White (1939) provided what is still the most complete account of its osteology, but his work may now be supplemented with observations based on a greater number of specimens and more recently described species. Vaughn (1966) first reported on *S. sanjuanensis*, and numerous complete specimens of this species (Berman et al., 1987b) contribute significant new information regarding the vertebral column of the genus. Olson (1979, 1980) described *S. grandis* from the Lower Permian of Texas and Oklahoma, and the smallest species of *Seymouria*, *S. agilis*, from Upper Permian sediments of the Chickasha Formation of Oklahoma.

Williston (1911a) and Watson (1918) estimated that the vertebral column of *S. baylorensis* contained 23 presacral vertebrae. White (1939) later accepted this count, even though at least one of his specimens had 24 presacrals. Vaughn (1966) reported at least 24 presacrals in *S. sanjuanensis* (NTM 1033), although he speculated that one more could have been present. CM 28596 and CM 28597 also show 24 presacrals for *S. sanjuanensis* (Berman et al., 1987b). The one known specimen of *S. agilis* also has a presacral count of 24. The number of presacrals of *S. grandis* is unknown. The presacral count for *Seymouria* appears to have remained 24 across an extended geologic interval.

White (1939) defined a cervical region and illustrated a distinct neck region for *S. baylorensis*, but Vaughn (1966) and Berman (pers. comm., 1986) argue that the position of the pectoral girdle precludes the presence of a distinct neck (see Williston, 1911a: Fig. 19; Berman et al., 1987b: Fig. 6D).

Atlas-axis complex. The atlas-axis complex is poorly known among most species of *Seymouria*, but well-preserved elements in *S. sanjuanensis* have recently been described by Berman et al. (1987b). The following description is based primarily on Berman et al. (1987b), with additional observations from other species of *Seymouria* added where warranted.

The atlantal intercentrum is a low, blocky wedge with a cup-shaped anterior face. Watson (1918) reported laterally directed processes in *S. baylorensis*, but they have not been observed in any other specimens, including the well-preserved specimens available to Berman et al.

An atlantal pleurocentrum is present only rarely (usually in mature forms), and probably existed as a cartilaginous element. It may have ossified relatively late in ontogeny (Berman et al., 1987b). It could not have reached the ventral aspect of the vertebral column, as the posterior face of the atlantal intercentrum and the anterior face of the axial intercentrum clearly abutted against one another (Berman et al., 1987b: Fig. 10).

The atlantal neural arch consists of paired halves that project posterodorsally to lie on the anterolateral surface of the large axial neural spine (Berman et al., 1987b: Fig. 10). The distal tips of the arches are somewhat pointed. Specimens of *S. baylorensis* indicate that the anterior zygapophyses are small and smoothly concave, whereas the posterior zygapophyseal facets are flatter and angled posteroventrally. Paired structures of the proatlas are not known for *Seymouria*, but the atlantal arch does exhibit paired articular facets for them.

The axial intercentrum is a separate ossification, similar to the atlantal intercentrum in shape. Parapophyseal articulations of the axis are directed posterolaterally. The axial pleurocentrum is a separate ossification as well. It has only a limited ventral exposure, largely excluded by the axial and third intercentra, giving it the shape of an inverted wedge. It bears paired articulations for the neural arch dorsally.

White (1939) described the axial neural arch as little different from that of the atlas; however, the spine that surmounts it is fused medially and is significantly larger than any other of the presacral series (Berman et al., 1987b). This large, blade-like structure angles forward between the paired halves of the atlantal arch. The anterior zygapophyses, which cradle the atlantal arches, are lightly constructed, with a distinct medial angulation, but the posterior pair are more robustly built and oriented much closer to the horizontal plane. The transverse processes of the axis are directed slightly posteriad, terminating in the diapophyseal articulation of the axis.

Dorsal vertebrae. The amphicoelous dorsal centra in *Seymouria* gradually increase in length and diameter posteriorly through the column. Most specimens possess a slightly flared lip at the central articulation. This lip is not always complete ventrally in more posterior vertebrae or in the large vertebrae of *S. grandis*, possibly due to the presence of strongly developed intercentra. Such regions are not marked by finished bone and were probably continued in cartilage (Figure 21A, C, E, and G). As with the atlantal pleurocentrum, ossification may have occurred relatively late in ontogeny. Those vertebrae with a finished ventral aspect show no evidence of a ventral ridge or keel. The pleurocentra of some *Seymouria* vertebrae bear accessory articulations at their dorsolateral edges (Figure 21A, E, and G). These angle down from the ventrolateral border of the neural canal on the anterior face of the centrum and articulated with faintly defined complementary structures on the posterior face of the preceding centrum. The anterior pair is braced dorsally by a thin, vertically oriented wedge that forms the lateral border of the neural canal.

Intercentra are present throughout the column as separate ossifications. None are ankylosed to the pleurocentra as in *Kotlassia* (Bystrow, 1944). They are concave anterodorsally and flattened ventrally, much more so in the posterior dorsal region. Their anteroposterior extent was such that they sometimes excluded the pleurocentra from ventral exposure. The intercentra of *Seymouria* were surely continued as cartilagenous elements dorsally, but the presence of accessory articular structures at the dorsal limits of at least some pleurocentra indicates that the extension must have tapered to a narrow wedge (Figures 20 and 21).

The neural canal is circular in outline and slightly larger in the anterior portion of the column. The pedicel is about the same length as the centrum, with only a slight emargination marking its anterior and posterior margins. Anteriorly the transverse processes are strongly developed, extending in a wing-like fashion far beyond the lateral limits of the neural arch. By the mid-dorsal region the lateral expansion of the neural arches becomes greater than that of the transverse processes, obscuring them in anterior view. The latter slant anteroventrally, their articular surfaces in the shape of an inverted teardrop, sometimes curved into a kidney-shaped articulation (Figure 20C).

Seymouria is noted for its conspicuously expanded neural arches. Those of the anteriormost vertebrae are not heavily swollen, but become progressively wider (Figures 20, 21) until the 10th vertebra, which has a transverse measure similar to those remaining. The anterior and posterior faces of the neural arches are marked by triangular depressions. The anterior depression is smaller and more clearly delimited than that posterior. Just lateral to the anterior depression the inner margins of the anterior zygapophyses are sometimes modified into semicircular, downward-facing, accessory articulations. They rested on a ridge at the ventral border of the posterior depression.

The lateral surfaces of some neural arches have conspicuous, anterodorsally oriented processes (Figure 21A and G). These are not as sharply defined as the mammillary processes found in *Eocaptorhinus* (Dilkes and Reisz, 1986), but probably served a similar function, for the attachment of M. spinalis dorsi and M. semispinalis. Just below and lateral to these processes, a distinct pit faces anterolaterally from in front of the posterior zygapophyses and appears to have provided for the attachment of the M. interarticulares.

The neural spines of the first few dorsals are taller and narrower than those of following vertebrae. Most of the neural spines are rounded and knobby, but occasional low, wedge-shaped spines provide for short regions of alternation of neural spine height. Additionally, some low-spined vertebrae exhibit furrows lateral to the neural spine, as in *Labidosaurus*. Fortuitously broken taller spines show that the lamellar bone of anterior and posterior walls was thicker than that of the lateral walls. No consistent pattern of alternation may be discerned, and the phenomenon seems to occur randomly in most specimens examined. However, a pattern of alternation is clearly present in the anterior portion of the column of *S. agilis*. The variability of neural spine structure is further expressed in their often bifid nature. Bifid spines are rarely divided evenly; often only a small spur emerges from one side of the neural spine. Curiously, the position of the smaller spur is sometimes seen to alternate from left to right in consecutive spines (clearly seen in MCZ 1083). A similar phenomenon has been reported in the trematopid amphibian *Anaconastes vesperus* (Berman et al., 1987a). Low-spined vertebrae are almost never bifid.

Sacral vertebrae and ribs. White (1939) reported one sacral vertebra in *S. baylorensis* and a modified caudal that he considered as a functional second sacral. *Seymouria*

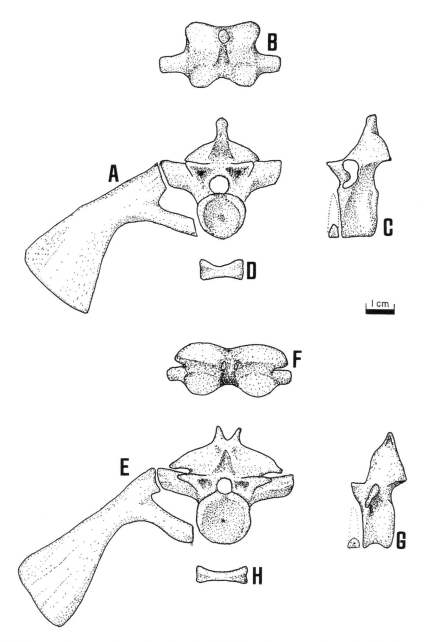

Figure 20. *Seymouria* sp. A-C, anterior, dorsal and left lateral aspects of anterior dorsal vertebra and associated right rib, approximately the 3rd or 4th presacral. D, anterior aspect of intercentrum associated with A-C. E-G, anterior, dorsal, and left lateral aspects of anterior dorsal vertebra and associated right rib slightly posterior to that illustrated in A- D, approximately the 6th or 7th presacral. H, anterior aspect of intercentrum associated with E-G. A-D based on UCLA VP 3152. E-H based on UCLA VP 570. Dotted lines indicate portion of pleurocentrum that was probably continued in cartilage.

sanjuanensis has only one sacral (Berman et al., 1987b), but *S. agilis* possesses three sacrals (Olson, 1980). Sacral structures are not known for *S. grandis*.

The sacral vertebra of *S. baylorensis* and *S. sanjuanensis* is massively constructed. In both, the centrum is blocky, almost flat ventrally, and dominated by the facets for the tubercular head of the sacral ribs. The pedicels are similarly dominated. The anterior zygapophyses are as widely spaced as those of the presacrals, and extend anteriorly to clear the rib facets. The posterior zygapophyses are much more closely approximated. The neural arches are not distinctly swollen, and the neural spines are often taller than those of dorsal vertebrae. The sacral intercentrum is a heavy element and longer anteroposteriorly than other intercentra. Like the sacral centrum, it is flat ventrally. A posteriorly directed facet articulates with the capitulum of the sacral rib.

The sacral rib is heavily built. The tubercular head is circular in cross-sectional outline. The capitular head, although smaller, is clearly separate, and has the outline of a horizontal oval. There is little or no constriction of the rib into a neck. The rib flares out into a wide articulation with the ilium. A strong iliosacral ligament was attached across the dorsal aspect of the rib (White, 1939).

The first sacral vertebra of *S. agilis* (UCLA VP 5329) conforms closely to the structure of the single sacral of other species of *Seymouria*. Its associated ribs are not well preserved and their abutment against the ilium cannot be observed. However, its distinctly narrower posterior zygapophyses clearly confirm its assignment to the sacral series. The neural arches and spines of the 2nd and 3rd sacrals are quite similar to those of the following caudals, and their associated ribs each make a firm, broad contact with the ilium (Olson, 1980: Fig. 8.3B).

Caudal vertebrae and ribs. Romer (1956) estimated the number of caudals in *Seymouria baylorensis* at about 40 to 60, whereas White (1939) estimated 40. The anterior caudal centra are similar in proportion to the sacral(s), except for the difference in costal attachment and a more rounded ventral aspect. More posterior caudal centra are spool-shaped. The first 4 to 6 neural arches are securely fused to their respective centra, but the remainder are not always firmly ankylosed to the centra. Neural arches are situated more directly above the centrum in the distal portion of the caudal series. The neural spines are tall, narrow, and blade-like, angling back sharply. There is no evidence of alternation of spine height. The first 6 caudal vertebrae have short recurved ribs, and the 1st haemal arch usually appears on the 6th caudal. White (1939) associated the variable position of the 1st haemal arch in *S. baylorensis* with the possibility of sexual dimorphism (also see Vaughn, 1966). Sexual dimorphism in *Seymouria* has been dismissed by Berman et al. (1987b) in a recent study of a number of complete vertebral columns; however, they did speculate on axial characters that could reflect on possible selection for increase in the size of the intromittent organ.

Presacral ribs. An atlantal rib was not found in any of the specimens examined, but Watson (1918) reported its presence, characterizing it as slender with clearly defined tubercular and capitular heads. The axial rib was short with a slight distal expansion, and the following 6 dorsal ribs were also expanded distally, the 5th to the greatest degree. Tubercular and capitular heads are clearly distinguishable in these anterior ribs (Figure 20). While separate, they are not distinguished by a distinct notch in the succeeding ribs, due to the relative lengthening of the capitular head and shortening of the

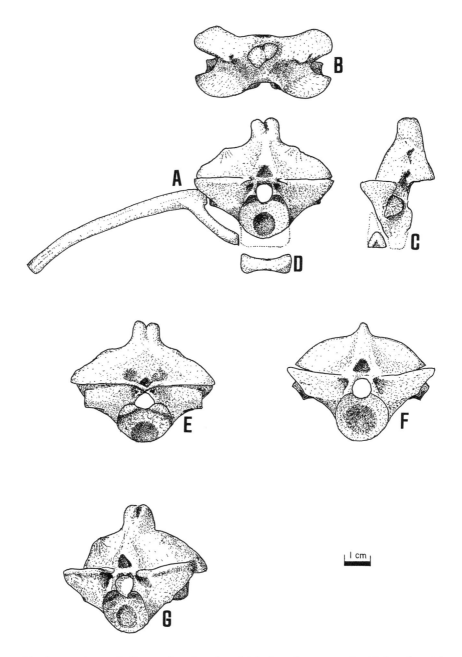

Figure 21. *Seymouria* sp. A-C, anterior, dorsal and left lateral aspects of mid-dorsal vertebra and associated right rib. D, anterior aspect of intercentrum associated with A-C. E, Posterior aspect of mid-dorsal vertebra. F, anterior aspect of a low-spined vertebra, probably *S. baylorensis*. G, oblique, anterolateral view of mid-dorsal vertebra to show accessory articular structures. A-E, and G based on UCLA VP 571. F based on UCLA VP uncatalogued, field number C-65-36. Dotted lines indicate portions of pleurocentrum and intercentrum that were probably continued in cartilage.

tubercular head. By the mid-dorsal region the tubercular head lies in line with the axis of the costal shaft, and the neck of the capitular head is oriented at a strong angle (Figures 20, 21).

DIADECTOMORPHA

Recently, the Diadectomorpha have come to be accepted as the probable sister group to amniotes, and thus warrant careful examination in regards to vertebral structure. The taxa included in the Diadectomorpha (sensu Heaton, 1980) still require careful study to elucidate their phylogenetic relationships. However, for the sake of convenience, they are presented here in the order considered to be from more primitive to more derived in Heaton's (1980) analysis of the group.

Limnoscelidae

The Limnoscelidae had long been considered to be representative of the most primitive "cotylosaurian" reptiles (Romer, 1944), but more recent work (Romer, 1966; Heaton, 1980) has removed them from that position. Heaton (1980) proposed that the limnoscelids are the most primitive grouping within the Diadectomorpha. Although the characters that he used have been challenged (Holmes, 1984; Smithson, 1985), a clear understanding of limnoscelid axial structure remains important to an understanding of the evolution of the vertebral column toward the amniote condition.

Limnoscelis paludis, the type species and best-known of the Limnoscelidae, is known from the extremely well-preserved holotype, YPM 811, as well as other materials. Less complete specimens have provided the basis for three other limnoscelid genera, none of which includes substantial vertebral materials. Romer (1952) described 16 posterior dorsals of *Limnosceloides dunkardensis*, none of which has preserved neural spines; and Langston (1966) referred to *Limnosceloides brachycoles* a few vertebrae that are extremely seymouriamorph in form. Lewis and Vaughn (1965) referred 7 fragmentary vertebrae to *Limnoscelops longifemur*. The vertebrae available to Carroll (1967) for recognition of *Limnostegyis relictus* were jumbled, and the neural spines preserved on only a few. The anterior half of the postcranium is preserved, but fragmentary, in *Romeriscus periallus* (Baird and Carroll, 1967).

Limnoscelis

The complete vertebral column in YPM 811 indicates that *Limnoscelis* possessed 26 presacral vertebrae. No attempt is made to subdivide the dorsal series.

Atlas-axis complex. The atlas-axis complex of YPM 811 is preserved on the inner surface of the interclavicle, making a ventral view impossible. The atlantal intercentrum is a large, separately ossified element. Due to the mode of preservation, the ventral extent of neither it nor the atlantal intercentrum can be determined. An atlantal rib is directed caudally from a process on the intercentrum. The atlantal centrum is not clearly visible, but can be seen to support paired neural arches. The arches cant posteriorly in a manner similar to the condition in *Seymouria*. A proatlas is not preserved, but probably did exist.

The axial rib is short and stout. The axial arch is not swollen, but is surmounted by an extremely large neural spine. The axial neural spine has a constricted cranial border that expands posteriorly, giving it a triangular outline in dorsal view and giving the convex lateral surfaces of the spine a noticeably anterolateral orientation. The dorsal half of the axial spine's posterior aspect bears a large circular pit. Beneath this, a median ridge separates two caudally directed fossae that are laterally delimited by thin lips of bone. The neural spine of the third dorsal was probably tall enough to accept muscular or ligamentous structures from the axial spine. The posterior expansion of the axial neural spine in *Limnoscelis* is greater than in *Seymouria* or any diadectomorph for which the structure is known.

Dorsal vertebrae. See Plate 3. Williston (1911b, 1912) provided a general outline of the dorsal vertebrae, reporting minimal change through the column. He described conspicuously swollen neural arches, similar neural spine proportions among vertebrae, and little change in the size of the centrum. However, YPM 811, CM 47653 (a new species of *Limnoscelis*), and materials from the Field Museum of Natural History reveal more variability than previously attributed to *Limnoscelis*.

Articulated materials indicate that the intercentra, though not quite as robust as those in *Seymouria*, probably had a substantial dorsal continuation in cartilage. The dorsal limit of such an extension is difficult to estimate, but no hyposphenes, episphenes, or complementary accessory articulations that might have limited their extent have been observed.

Centra are amphicoelous and notochordal. The anteriormost dorsal centra are more squatly cast and spool-shaped than those posterior. They have a smaller diameter and are 20-25% longer anteroposteriorly. The proportions illustrated by Heaton (1980) are more typical of the posterior region of the column. Farther back in the column, the centra approach a disc-like shape and have anterior and posterior lips. Bevelling for the intercentra is not strong posteriorly. As in the posterior centra in *Labidosaurus*, paired ventral ridges delimit a deep midventral depression by the middle of the column.

The neural arch pedicels and the diapophyseal portion of the transverse process are proportionately longer anteriorly in the column. In anterior or posterior aspect, the transverse processes are strongly flared through the first two-thirds of the column. Laterally, their outline is that of an inverted teardrop with a gently sigmoidal curve.

The dorsal neural arches are expanded throughout (Plate 3A and E) but less so in the anterior region of the column. Those anterior (Plate 3A) are also longer anteroposteriorly. The 3rd neural spine was stout, as was the 5th; both are round at the base, about 9 mm. in diameter. Such tall spines have a sharp median ridge on their anterior and posterior faces (Plate 3D and E), much like that in *Diadectes* (Figure 23), but with the tip more rugose and a bit wider. The 4th neural spine is a narrow, low ridge, about 3.5x12 mm. CM 47653 includes an isolated vertebra (Plate 3B) very close in structure to the 4th of YPM 811. Its spine is long and low, and the arch bears laterally placed furrows like the low-spined vertebrae of a number of other taxa. A string of vertebrae contained in FMNH UR306 has similarly low spines at positions that would correspond to approximately the 6th and 9th or 7th and 10th dorsals of YPM 811. The 6th dorsal spine in the holotype is perfectly preserved, and clearly a tall type, measuring 16.7 mm. in height and 11 mm. in basal diameter. The 7th and 10th neural spines are also tall, but distinctly longer and narrower than the 6th or 9th. Although the patterns of alternation are not exactly identical between the various specimens, some form of alternation of neural spine structure is consistently present in the anterior region of the axial column.

The zygapophyseal planes of the anteriormost vertebrae are strongly tilted in *Limnoscelis* (Plate 3B). Anteriorly they are tilted inward at an angle of approximately 30 degrees, and posteriorad at about 25 degrees. It appears that little lateral movement would have been possible at these articulations. Instead, flexion in a manner to raise the anterior end of the column, as well as a small amount of rotation, would have been more likely. The 3rd to 7th dorsals are positioned between the shoulder girdle and like *Seymouria, Limnoscelis* had no truly functional neck. Presumably the heavy, spatulate ribs provided a strong connection between the axial column and the limb girdle. This, combined with alternation in neural spine height or structure, may have coupled the motion of the pectoral girdle with that of the vertebral column.

Presacrals 12 to 16 show clear alternation of neural spine height, with 13 and 15 bearing almost no spine at all. The neural spines of the 12th, 14th, and 16th dorsals are constructed like the tall types situated more anteriorly. The 17th and 19th are also tall types, and though 18 has a spine that is also relatively tall, it is relatively quite narrow, only about 5 mm. in width. The zygapophyses in this region are no longer canted posteriorly, but their articular planes still tilt inward at an angle of 15 degrees.

The most posterior dorsals continue to display gradual changes in structure of the centra, pedicels, and transverse processes. Almost all have tall neural spines, though number 22 of the holotype may have been a low-spined type. (The spine is missing, but a long, narrow break mark remains.) Anterior- and posterior-facing median ridges of the neural spines are not distinct in the type specimen, but disarticulated elements of CM 47653 display them clearly. The tip of the last presacral is visibly expanded into a strongly bifid structure. In some vertebrae, rough prominences extend from the lateral surface of the lower portion of the neural spine. They often reach to the anterolateral limit of the neural arch, projecting forward through much of their length. Presumably they marked the attachment of well-developed M. spinalis dorsi and M. semispinalis. The well-defined anterior and posterior median ridges of the neural spines indicate the presence of paired Mm. interspinales.

Posteriorly, the zygapophyses retain an inward tilt of 15 to 20 degrees. While there is no conspicuous alternation of spine height, interspinous distances are great enough, and zygapophyses tilted enough, to allow at least a small amount of horizontal flexure.

Sacral vertebrae and ribs. Limnoscelis possesses two sacral vertebrae. The first is so much more massive that Williston (1911b) believed only one sacral to be present and Romer (1944) stated that *Limnoscelis* was "...transitional in development between a one-ribbed and two-ribbed condition." The ventral aspect of the sacral complex is not accessible, and the neural spines are broken off. As in many other forms with swollen neural arches, the anterior zygapophyses of the first sacral are widely spaced, but the posterior pair is much closer together, and the arch is not visibly expanded. The arch of the second sacral is likewise narrow. The ribs of the first sacral are much larger than those of the second, and their distal ends widened and downturned for a broad and nearly vertical contact with the ilium. The second pair of sacral ribs articulates with the ilium in a restricted manner, though they may have provided additional support via ligamentous attachments.

Caudal vertebrae and ribs. There were approximately 60 caudals in YPM 811, the first few strongly resembling the second sacral. The neural arches are not swollen, and the transverse processes are oriented directly laterad. Caudal centra decrease in length very slowly. Intercentra are present, and chevrons attached beginning with the 3rd caudal. Neural spines exhibit no evidence of alternation. They are tall and blade-like in the anteriormost region of the tail, but decrease in height quickly. Sharply recurved ribs are present on the first 10 or 11 caudals (Plate 3J).

Presacral ribs. The dorsal ribs are holocephalous throughout, but Romer (1944) speculated that their proximal ends may have had cartilaginous caps that permitted the passage of the vertebral artery between the capitular and tubercular heads. As mentioned previously, the distal ends of the 3rd to 7th ribs were broadly spatulate, for muscular attachment to the pectoral girdle (Plate 3I). The next few ribs are flattened distally, but are not particularly broad. The 10th rib is the longest of the series. Posteriorly they become shorter, and subcircular in cross-section.

Tseajaiidae: *Tseajaia*

Tseajaia campi, the only described member of the Tseajaiidae, was first described by Vaughn (1964), in a brief preliminary account wherein he speculated that it could be a morphological link between seymouriamorphs and diadectomorphs. Romer (1966) later included it with the Seymouriamorpha. Moss (1972) followed Romer's assignment to the seymouriamorphs, and proposed that a variety of cranial characters demonstrated a close relationship between *Seymouria, Tseajaia,* and *Diadectes*. Heaton (1980) assigned *Tseajaia* to the Diadectomorpha, indicating that it shared only primitive characters with seymouriamorphs, of which he considered the structure of the vertebral column to be one of the most striking. Despite Heaton's reservations about the phylogenetic utility of the vertebral column, its extreme similarity to that of *Seymouria* points to important functional similarities and makes their immediate comparison appropriate.

Atlas-axis complex. The atlas-axis complex in *Tseajaia* has been described and figured by Moss (1972); however, reinterpretation of these structures is necessitated by the description of well-preserved materials of the atlas-axis complex in *Seymouria* by Berman Reisz and Eberth (1987b). Re-examination of the type specimen, UCMP V4225/59012, indicates that, contrary to Moss' (1972) reconstruction, the axial neural arch and neural spine in *Tseajaia* were almost identical in construction to that in *Seymouria sanjuanensis*. Further, whereas the axial centrum is somewhat more robustly constructed than that of *S. sanjuanensis*, it is similarly wedge-shaped ventrally.

As preserved in UCMP V4225/59012, the atlantal intercentrum and axial intercentrum abut against one another quite closely, and it seems likely that they had a similar orientation in life. Moss (1972) felt that the atlantal neural spine was low and poorly developed, but proper articulation of the posterior zygapophyses of the atlas with the axis would indicate that the spine is somewhat longer than depicted by his reconstruction, with a configuration very much like that in *S. sanjuanensis* (Berman et

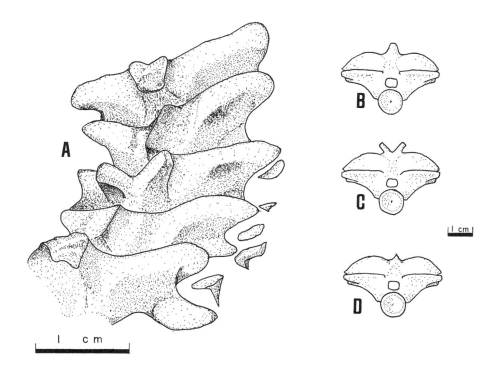

Figure 22. UCMP V41225/59012, *Tseajaia campi*. A, anterodorsal view of presacral vertebrae 5 through 9 showing alternation in neural spine structure and the strongly bifid condition of presacral number 7. B-D, reconstructions in anterior aspect of vertebrae of *Tseajaia*, showing tall, bifid, and low-type neural spines. The reconstructions of B-D are purposely simplified, due to the inaccessible nature of much of the column in UCMP V41225/59012.

al., 1987b: Fig. 10). Further, the position of the transverse process argues for a more dorsally placed atlantal pleurocentrum than that figured by Moss (1972: Fig. 6A and B), indicating that it was more likely wedged between the atlantal and axial intercentra with no exposure on the ventral surface of the column, much as in *S. sanjuanensis*. Moss (1972) indicated the possible presence of proatlantal fragments, but their preservation is so marginal as to make their identification equivocal.

Dorsal vertebrae. All complete *Tseajaia* specimens examined indicate a presacral count of 26. The dorsal vertebrae are virtually indistinguishable from those of *Seymouria*. The similarity is so striking that isolated vertebral material assigned to *Seymouria* might actually belong to *Tseajaia*. Little description needs to be added, save for the few minor differences from *Seymouria*. The centra measure only about 25% of the width of the neural arch (Heaton, 1980), which is somewhat less than in *Seymouria*. In contrast to *Seymouria*, some posterior dorsal centra have midventral ridges. Accessory articular structures of the pleurocentra are not known, but because the specimens are articulated this feature cannot be confidently assessed. The intercentra were surely continued dorsally in cartilage, and it appears they were not restricted dorsally as in some *Seymouria* vertebrae. The transverse processes in *Tseajaia* are more stoutly buttressed than those of *Seymouria* by means of a ventromedially directed web that reaches to the midheight of the lateral surface of the centrum (Figure 22B-D).

Tseajaia exhibits more variability in neural spine morphology than does *Seymouria*, and alternation of spine height is clearly visible in UCMP V4225/59012. Low-spined vertebrae are much like those in *Seymouria*, with furrows lateral to the neural spine even more clearly developed. Tall neural spines take on a number of configurations, including high and rounded, bifid, and in at least one instance, strongly forked (Figure 22). As in some specimens of *Seymouria*, alternation of neural spine height is located in the anterior portion of the holotype. CM 38042 and CM 38033, from the Lower Permian Cutler Formation of New Mexico, do not display alternation of spine height, but the former does show a pattern of wide or lunate-shaped spines alternating with narrow, more anteroposteriorly extended spines in presacrals 5, 6, 9 to 15. Adjacent spines are too badly damaged for description. The holotype exhibits alternation of spine height more posteriorly in presacrals 16 to 20 as well.

Sacral vertebrae and ribs. *Tseajaia* possesses two sacral vertebrae. The first is more heavily constructed, and its associated ribs account for the bulk of the iliosacral articulation. The shape and proportions of the first sacral vertebra and ribs are comparable to those in *Seymouria,* but the ribs are deflected more sharply posteriad to their ilial articulation. Sacral neural spines are not well preserved in any of the specimens of *Tseajaia* studied. The second sacral vertebra is built like those of the caudal series and is quite slender relative to the first. It extends directly to its contact with the ilium without touching the first sacral rib.

Caudal vertebrae and ribs. Long thought to be lost, the block containing the tail of the type specimen of *Tseajaia campi* has recently been found, but was not available for study. On the basis of photographs by Hilton (1942), Moss (1972) judged the tail to be about as long as the presacral column. The caudal elements seem to be little different from those in *Seymouria*.

Presacral ribs. *Tseajaia* appears to have ribs associated with all of the presacral vertebrae. They are very much like those in *Seymouria* and *Diadectes* in most respects, though the distal expansion of anterior dorsal ribs is not as marked as those in *Diadectes*.

Diadectidae

Numerous genera and species have been assigned to the Diadectidae. Romer (1944) and Olson (1947), however, have clarified considerably the picture of the North American representatives, recognizing only three genera: *Desmatodon, Diadectes*, and *Diasparactus*. The genus *Diadectes* itself contains a number of species, but their specific distinction depends primarily on cranial characters (Case, 1911; Watson, 1916; Romer, 1944; Olson, 1947, 1965, 1966; Moss, 1972), the details of which fall beyond the scope of this study. Descriptions provided by Case (1911), Williston (1925), and Romer (1944,

1956, 1966) have provided an understanding of its postcranial skeleton, and Olson's (1936) work on the axial osteology and musculature of *Diadectes* remains the best account available. Heaton (1980) provided an illustration of one dorsal vertebra of *Diadectes*, but its extremely narrow lateral width appears to be in error. Whereas cranial characters define the different species, Olson (1947) has noted that the postcranium in *Diadectes* has remained "...decidedly stable throughout the genus and few differences of even specific value can be demonstrated."

Considering the close relationship between *Seymouria*, *Tseajaia*, and *Diadectes*, it is surprising that no specimens of *Diadectes* are known to show alternation of neural spine height. One specimen (CM 38036) that may be assigned to the Diadectidae with a fair degree of confidence does display this phenomenon. Because the postcranial skeleton in *Diadectes* is reasonably well known, a complete redescription will not be pursued here. The basic structures of diadectid vertebrae are illustrated in Figure 23A-C. The configuration of CM 47653 and the atlas-axis complex will be presented in reference to the discussion already conveyed in this section.

Atlas-axis complex. Elements of the atlas-axis complex have been figured by Olson (1936) and Evans (1939). However, careful examination of the well-preserved atlas and axis vertebrae in FMNH UR27 warrants the reinterpretation of certain portions of the complex. Romer (1944) noted that elements of the atlas-axis complex are often fused in mature specimens. Although it is difficult to interpret the pattern of sutures, Romer (1944) felt that the ventral limit of the first vertebral segment may consist of the fused atlantal and axial intercentra, completely excluding the atlantal pleurocentrum from exposure on the ventral surface of the axial column. The atlantal intercentrum is thick and "U- shaped" in end view. It bore well-developed parapophyses for the atlantal rib. Immediately posterior to it, a single, thick, blocky element with stout costal processes may be identified as the axial intercentrum. As there is no room for intervening elements, the atlantal pleurocentrum could not have reached the ventral margin of the vertebral column. The atlantal pleurocentrum is represented by a thin disc of bone that has fused to the dorsal aspect of the axial intercentrum. Evans (1939) mistakenly interpreted the fused atlantal pleurocentrum as a neural spine of the axial intercentrum.

If the above interpretation of the atlantal pleurocentrum is accepted, the atlantal neural arches would have had a more posterodorsally placed point of articulation than has been illustrated for *Diadectes* previously (Olson, 1936; Evans, 1939). The atlantal neural arches each pass a tapered spine just behind an anteroventral process of the axial neural spine to a position lateral to the axial spine. Although the atlantal neural spines are not as long as those in *Seymouria* or *Tseajaia*, the relative positions are much the same. The proatlas, partially preserved in FMNH UR27 is thick and robustly constructed.

The axial centrum was well ossified and was constricted between the axial and third intercentra, though not to the degree in *Seymouria* (Berman et al., 1987b). It did reach to the ventral border of the vertebral column. The axial neural arch bears a diapophysis that is turned in a sharply posteroventral direction for the distinctly bicipital axial rib. The axial neural spine is large and triangular in lateral view. A hooked process marks the anteroventral corner of the spine. Evans (1939) did not restore this portion of the spine, but Olson (1936) interpreted it as the point of origin for the M. obliquis capitis superior. In light of the position of the atlantal neural arch proposed here, Olson's (1936) interpretation is considered more accurate than that of Evans (1939). While the axial neural spine is substantial in size, it is not as anteroposteriorly elongate as those in *Seymouria* or captorhinid reptiles. The anterior edge of the spine is sharp, but the spine thickens to a broad posterior margin. The posterior margin is marked by a series of furrows, much as in *Captorhinus*. The lateral surface of the spine provided ample area for attachment of the M. obliquis capitis magnus. The tip of the spine is expanded into a rugose, bifid knob, presumably for attachment of paired M. interspinalis and firm ligamental attachments.

CM 38036. In June of 1979, D. S Berman of the Carnegie Museum of Natural History and field party collected the greater part of what appears to be a juvenile diadectid (CM 38036) from near Arroyo de Agua in northern New Mexico. Much of the cranium is badly fragmented, and although specific referral is not possible, the diadectid nature of the specimen seems certain. The form of the humerus is extremely close to that of *Diadectes*. The proportions and shape of the shaft and adductor crest of the associated femur are also suggestive of that genus; but, as would be expected for a juvenile, the proximal and distal ends are not as completely ossified as in adult diadectids. The teeth preserved in CM 38036 are transversely expanded, and the incisors are characteristically procumbent. Though they lack the wear facets and clearly developed cusps seen in *Diadectes*, the differences in dental structure may not be significant. Vaughn (1969, 1972), has shown that the juvenile teeth of the primitive diadectid *Desmatodon* are not as strongly cuspate as those of adults. *Diasparactus zenos* (Case, 1910; Case and Williston, 1913) was described by Case (1910) as having smaller centra, relative to the spread of the neural arches, than other diadectids. *Diasparactus* is quite small for a diadectid, but has disproportionately long neural spines on the posterior dorsal vertebrae. The proportions of the centra are about the same size as in CM 38036, but the neural spines of the latter are not nearly so long.

The atlas and axis of CM 38036 are missing, and the 4th and 5th presacral are preserved beneath the scapulocoracoid. However, the 3rd dorsal and the remainder of the dorsal series are preserved. Matrix and ribs obscure from clear lateral view much of the vertebral column. The neural spine of the 3rd dorsal is broken off, but the break mark indicates a rather narrowly triangular spine in cross-section. In dorsal aspect the neural arches of CM 38036 are very similar to those in *Seymouria* and *Tseajaia* (Figure 21D and E). Their zygapophyses are displaced far laterally and the arches extremely swollen. As far as can be determined, the broad neural arches were supported by flared buttresses typical of diadectids (Figure 23A-C). The centra are proportioned as those in *Diadectes* (Figure 23A-C), and are not quite as small as those in *Tseajaia*.

As in *Tseajaia* and some *Seymouria*, alternation of neural spine height is evident in the anterior half of the axial column in CM 38036 (Figure 23D). Presacrals 6 through 14 alternate in height, though the pattern is imperfect in that both the 9th and 10th possess tall-type spines. The low-type neural spines of CM 38036 are constructed much like those in *Captorhinus*. Although they are more similar in size to those in *Labidosaurus*, the neural arches of CM 38036 do not display troughs lateral to spines.

The tall-type spines of CM 38306 are even more heavily built than those in *Seymouria, Tseajaia*, and captorhinids, and more like typical diadectid neural spines

Vertebral Structure in Permo-Carboniferous Tetrapods

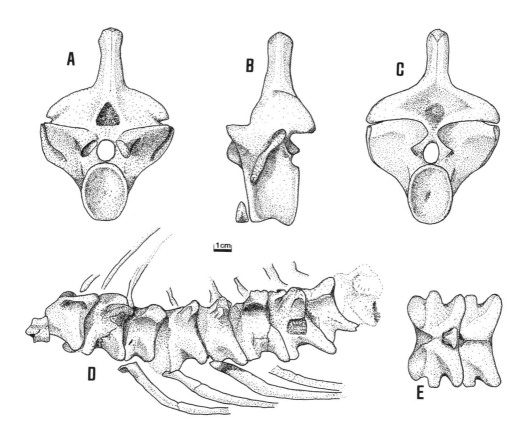

Figure 23. Vertebrae of diadectids. A-C, *Diadectes* sp. mid-dorsal vertebra; anterior, left lateral, and posterior aspects. A-C based on UCLA VP 344, UCLA VP 2962, UCLA VP 3725, and UCLA VP 2335. D, CM 38036, a juvenile diadectid showing alternation of neural spine height in presacrals six through fourteen. E, reconstruction in dorsal aspect of vertebrae with high and low-type neural spines based on CM 38306.

(Figure 23A-C). They are about three times taller and two to three times wider than the adjacent low-type spines (Figure 23D and E). A sharp median ridge separates a pair of deep concavities on the anterior and posterior surfaces of the spines that undoubtedly provided attachments for paired Mm. interspinales. The tips of the spines are rugose and slightly flared, indicating that ligamental or fascial attachments might have been well developed. Laterally, distinct ridges curved forward, presumably providing attachment for the Mm. spinalis dorsi and Mm. semispinalis. These ridges or concavities do not occur on the low-type neural spines of CM 38036.

Dorsals 17 to 23 are all quite robust and do not show any evidence of alternation in structure. They are even blockier than those of the more anterior tall-spined vertebrae. The posterior neural spines do not have a clearly defined sagittal ridge, and it is not possible to tell if the M. interspinalis was paired or a single mass. It is also difficult to determine distinct points of attachment for the M. spinalis dorsi or M. semispinalis. The 20th dorsal, possessing the most lightly constructed of the posterior dorsal neural spines, shows only a faint indication of such an attachment. Although the posterior neural spines of CM 38036 are heavily built, they are not nearly as tall as those of *Diasparactus*, which are as tall as the centrum and neural arch combined. It does not appear that CM 38036 is referable to *Diasparactus*.

Alternation can be demonstrated with certainty for only one specimen that seems to be referable to the Diadectidae but, as in *Tseajaia* and some specimens of *Seymouria*, the phenomenon is concentrated anteriorly. Its presence in a juvenile form makes interpretation difficult. Because of the clean, finished nature of the bone surface in regions of alternation, it does not seem likely that a radical remodeling of vertebrae and epaxial musculature would have occurred through ontogeny. At the very least, CM 38036 provides a diadectid representation of alternation of neural spine height, and does so in a pattern quite like those of the previously discussed diadectomorphs.

SEYMOURIAMORPHA AND DIADECTOMORPHA: SUMMARY

The close relationship of the seymouriamorphs and diadectomorphs has considerable support (Olson, 1947, 1965, 1966; Vaughn, 1966; Moss, 1972; Heaton, 1980; Panchen, in Panchen and Smithson, 1988), although Smithson (in Panchen and Smithson, 1988) has expressed a dissenting view. Although considered by some to be a reflection of shared primitive character (see Heaton, 1980), nonetheless their similarity in vertebral structure serves to reinforce this general conclusion. Most seymouriamorphs and diadectomorphs possess features adapted for structural stability within the vertebral column. Heavy neural arches, accessory articulations, and potentially strong ligamentous connections are more obvious than in the captorhinids. Additionally, the presence of characteristics that would allow for columnar dorsiflexion seems to have been overlooked, especially for *Limnoscelis*.

The pattern of alternation is irregular and unpredictable in most examples of *Seymouria*, suggesting that the more primitive function of neural spine variability may have been related to a muscular contribution to the axial stability provided by the features outlined above. In *S. agilis* and relatively more derived forms such as *Tseajaia* and at least one diadectid, the alternation becomes a more clearly defined phenomenon and is expressed close to the pectoral girdle. As alternation is not present in all forms of these closely related groups, the conclusion that it has developed in parallel to the condition in captorhinids must be considered. Although the concentration of alternation near the pectoral girdle is different from that interpreted as primitive in the captorhinids, such a pattern suggests that its functional relation to the limb girdles must have been very important.

4

THE MICROSAURIA AND OTHER "LEPOSPONDYLS"

Traditionally, the amphibian order Microsauria has been placed in the subclass Lepospondyli (Romer, 1947, 1950, 1966; Baird, 1965), a group named for its vertebral structure. Lepospondylous vertebrae are characterized by their spool-shaped centra that, presumably, ossified as a single unit. Usually included with the Microsauria in this subclass are the snake-like Aistopoda and the mostly aquatic, salamander-like Nectridia.

The Lepospondyli is almost surely an artificial grouping. Gregory et al. (1952) and Carroll and Gaskill (1978) noted significant differences in the vertebral structure, one of the defining characters of the subclass, of its constituent groups, and Panchen (1980) pointed out that none of them appeared to be closely related. However, the order Microsauria has been demonstrated to be a valid taxonomic group by Romer (1950) and Carroll and Gaskill (1978). Carroll and Gaskill have provided extensive historical, morphological, and phylogenetic analyses of all the species assigned to the order. While an evaluation of the vertebral structure of all microsaurs is not the purpose of this study, a number of them display alternation of neural spine height, and a description of these taxa is warranted here.

MORPHOLOGICAL CHARACTERISTICS OF THE MICROSAURIA

The Microsauria is a difficult group to characterize morphologically, and extreme variability in the presacral count among its various members exemplifies these difficulties. As noted above, the presence of lepospondylous vertebrae has been a character by which its members have been grouped; but many have been shown to possess discrete intercentra (Vaughn, 1972; Carroll and Gaskill, 1978). Carroll (1989) has speculated on the developmental aspects of lepospondylous vertebrae; however, ontogenetically defined criteria remain problematic in the interpretation of fossil forms.

The first attempt to define the microsaurs morphologically was by Romer (1950), and more recently Carroll and Gaskill (1978) have listed a number of criteria characterizing the group. Microsaurs: (1) lack an otic notch, (2) lack the labyrinthodont pattern of palatal fangs, (3) lack the typically reptilian transverse flange of the pterygoid, (4) possess a unique arrangement of the first presacral vertebra and its articulation with the occipital condyle, and (5) have a distinctive structure of the trunk and caudal vertebrae. As the first three criteria are based on negative evidence, the importance of the vertebral column in description of the Microsauria assumes greater significance.

The first presacral vertebra of most microsaurs is a single unit, as opposed to the distinct ossifications of the atlas-axis complex in other Paleozoic tetrapods (Carroll, 1968); a broad, "strap-shaped" articular surface formed by the basioccipital and exoccipitals facilitated articulation with the skull. More posteriorly, the trunk and caudal vertebrae have been characterized by their lepospondylous nature. However, as with other groups, the alternation in structure of the neural arches and spines was largely ignored until the work of Carroll (1968), Vaughn (1972), and Carroll and Gaskill (1978) pointed out the pervasiveness of this phenomenon among microsaurs.

PANTYLIDAE: *PANTYLUS CORDATUS*

While other forms have recently been assigned to the Pantylidae (Berman et al., 1988), only *Pantylus* has vertebral material of sufficient quality and preservation to confirm the presence of alternation of neural spine height. The monospecific genus *Pantylus*, known from the Early Permian redbeds of Texas, is among the most thoroughly described of all microsaurs. Although early accounts (Cope, 1881, 1892, 1896; Case, 1911; Mehl, 1912; Williston, 1916) contributed little to an understanding of its postcranial anatomy, the works of Romer (1969) and Carroll (1968) have since provided a complete description of its morphology.

Carroll (1968) estimated a presacral count of 24 in *Pantylus*. He observed that *Pantylus* probably had only one sacral vertebra, but its structural details and those of and its associated ribs are not visible in any of the specimens available for study.

Atlanto-axial vertebra. The first vertebral element of the presacral column in *Pantylus* is unusual, even among microsaurs. The centrum is extremely long, exhibiting articular processes for two successive ribs, indicating that the first presacral incorporates fused structures homologous to the atlas-axis complex of other Paleozoic tetrapods (Carroll, 1968). Anteriorly, a distinct "odontoid" process articulated in a ball and socket joint with the basioccipital and exoccipitals. The region just behind the odontoid is wider than the anterior and posterior limits of the centrum. Posteroventrally, the atlanto-axial centrum is tightly articulated with a large intercentral element, presumably that of the next dorsal vertebra (Carroll, 1968; Carroll and Gaskill, 1978).

There is no distinct constriction of the pedicel of the neural arch in the first vertebra, but there is a massive neural spine. Heavily built, squared off, and blocky (Carroll and Gaskill, 1978: Figs. 115A and L, 116C), it is similar in proportion to the axial spine of many anthracosaurian amphibians and primitive reptiles. It is somewhat longer anteroposteriorly than it is wide, and almost as wide as it is high. The tip of the spine in MCZ 3302 is chipped, and its dorsal extent may be somewhat greater than that illustrated by Carroll (1968). The dorsal margin of the spine angles anteroventrally in a manner similar to that of many captorhinids. Anterior to the large neural spine, an accessory neural arch, much like the atlantal arch in primitive reptiles, bears costal processes, but lacks a neural spine and articulations for proatlantal structures.

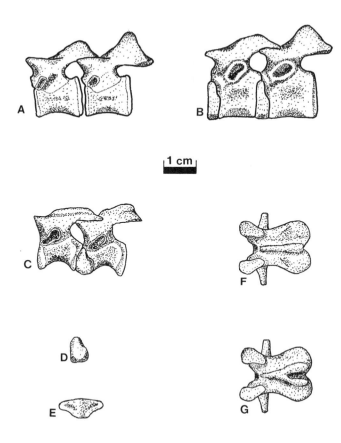

Figure 24. Microsaurian vertebrae. A, *Pantylus cordatus*, left lateral aspect of dorsal vertebrae in articulation. Based on MCZ 3302. B, *Ostodolepis brevispinatus*, left lateral aspect of dorsal vertebrae in articulation. Based on FMNH UC680. C, *Trihecaton howardinus*, left lateral view of dorsal vertebrae in articulation. D-E, *T. howardinus*, intercentrum in left lateral and ventral aspects. F, *T. howardinus*, low-spined mid-dorsal vertebra in dorsal aspect; anterior to the left. G, *T. howardinus*, tall-spined mid-dorsal vertebra; view as in F. C-G based on UCLA VP 1743.

Dorsal vertebrae. The dorsal centra are amphicoelous, single ossifications. There is no evidence of intercentral elements between consecutive dorsal centra. A "V-shaped" suture separates the centra and their associated neural arches (Figure 24A). The neural arches are not as conspicuously expanded as those of seymouriamorphs, diadectomorphs, or captorhinids; their lateral limits do not extend beyond the transverse processes. The arches are, however, relatively wider than those of protorothyridids and most pelycosaurs, retaining a convex outline in end view. The transverse processes extend outward to the level of the lateral extent of the centra.

Alternation of neural spine height is clearly expressed in the anterior portion of the column in MCZ 3302 (Figure 24A). Following the heavily constructed atlanto-axial vertebra, the 2nd and 4th vertebrae bear negligibly developed neural spines, similar in their degree of development to those of some specimens of *Captorhinus* (Figure 7B). The neural spine of the 6th vertebra is taller than those of the 2nd and 4th, but still shorter than those directly adjacent to it. Presacrals 3, 5, and 7 possess taller, conical neural spines. Their tips are slightly rugose, approaching a bifid orientation at their apex.

Carroll (1968) assumed that alternation continued throughout the length of the presacral column. However, other taxa indicate that a perfectly regular pattern of alternation is rare. The break mark outlining the remains of the 8th neural spine of MCZ 3302, though narrower than the base of the 7th, is not as distinctly different as those of the anteriormost region of the column. This condition, and the smaller relative difference between the 6th neural spine and the two adjacent to it, may indicate that alternation faded more quickly than Carroll assumed. Confirmation of either possibility awaits the discovery of better material. Sacral structures were not accessible in any of the specimens examined.

Caudal vertebrae and ribs. *Pantylus* apparently had a very short tail; Carroll (1968) estimated that the caudal series had only about 12 segments. Proximal caudal centra were probably single ossifications with no associated intercentral elements. However, haemal arches, almost always derived from intercentral elements, are present from the 4th or 5th caudal back. The haemal arches are much like those of *Protocaptorhinus* (Figure 16B). Ventrally, the caudal pleurocentra bear paired ridges.

The first 5 or 6 caudals have transverse processes that articulated with the tubercular processes of bicipital ribs. The capitular processes articulated very close to the anteroventral limit of the centrum. Neither caudal ribs nor transverse processes are evident posterior to the 6th caudal vertebra. The neural arches of the caudal vertebrae are firmly attached to the centra in mature forms (Carroll, 1968), though faint indications of the sutures remain.

Pantylus is the only tetrapod known to display alternation of neural spine height in the tail region. The first caudal has a spine taller than that of the second, and alternation continues throughout the caudal series. The absolute and relative differences in height of adjacent neural spines are, however, not as great as those of the anteriormost presacral vertebrae, and no difference in the construction of the associated neural arches can be detected. Additionally, the neural spines of the caudal vertebrae do not differ significantly in midline length. Although alternation appears to fade in the mid-dorsal region of the vertebral column, its presence in the caudal series may be suggestive of its presence near the pelvic girdle, a condition reminiscent of that usually seen in *Captorhinus*.

Presacral ribs. MCZ 3302 distinctly reveals the articulation of two successive ribs with the first vertebral element. They, and the succeeding four ribs, are clearly bicipital, with slightly expanded distal ends. The articular heads of those ribs located more posteriorly remain separate, but not as obviously as the first five. Carroll (1968) noted that the ribs extend far laterally, and proposed that *Pantylus* had a relatively wide trunk. The ribs shorten through the posterior third of the presacral column.

OSTODOLEPIDAE

All of the species currently assigned to the Ostodolepidae exhibit alternation of neural spine height. The modes of this expression are variable between species however, and they are described in turn.

Ostodolepis brevispinatus

Ostodolepis brevispinatus as currently defined by Carroll and Gaskill (1978), is known from only one specimen (FMNH UR680). Williston (1913) originally assigned the name *Ostodolepis* to an articulated series of seven dorsal vertebrae, ribs, and associated scales. He later thought that it might be synonymous with *Pantylus*, but retained the name *Ostodolepis* with some uncertainty (Williston, 1916). Case (1929) attributed a nearly complete skeleton to *O. brevispinatus*, but Carroll and Gaskill (1978) have pointed out differences in the structure of the neural spines between it and the type specimen, assigning Case's specimen to a new genus, *Pelodosotis*.

Carroll and Gaskill (1978) admitted that generic distinction based on only one string of seven vertebrae presents some difficulty, but retained the genus *Ostodolepis* and the familial name Ostodolepidae for the sake of convenience. They characterized the Ostodolepidae as a terrestrial, burrowing group; but applying such conclusions must be tentative given the small portion of the skeleton available for analysis. Although the paucity of material might also bring the microsaurian affinities of *Ostodolepis* into question, the scales associated with the vertebrae appear to be characteristic of the order.

Vertebrae. The centra in *Ostodolepis* (Figure 24B) are notochordal and deeply amphicoelous. They are approximately 50% taller than they are wide, and slightly shorter than they are high. Dimensions of the centra do not change appreciably through the short string. The anterior and posterior faces articulate only at their dorsalmost margins, otherwise separated by well-developed intercentra that are embayed posteriorly for reception of the capitulum of the rib. Dorsally, the intercentra extend nearly to the level of the anteroventrally angled transverse processes.

The zygapophyseal facets are only slightly tilted. The neural arches are low, but do not reach out beyond the lateral limit of the transverse processes. In lateral view, the neural spines are all subtriangular in shape, but superimposed on their outline is a moderate alternation in height. The first (interpreted as the seventh presacral by Carroll and Gaskill, 1978) and second of the series both have tall-type spines, the third is low, and alternation follows back to the end of the series. Tall-type spines are conical, but rounded at the tip, providing more anteriorly and posteriorly directed surface area for Mm. interspinales. Low-type neural spines are less than half the height of adjacent tall-type spines, and their associated neural arches are somewhat more convex. Low-type spines are much narrower transversely and longer anteroposteriorly.

Ribs. The ribs associated with FMNH UR680 are all clearly bicipital. Although most of the distal shafts are missing, they can be seen to be essentially cylindrical, except for the two most anterior. The first two ribs of the holotype are just slightly expanded distally, but not to the degree seen among captorhinids, seymouriamorphs, or diadectomorphs. The distal widening of the anteriormost ribs may indicate that the holotypic string may have been from a region close to the pectoral girdle.

Pelodostis elongatum

Carroll and Gaskill (1978) based the genus *Pelodosotis* on University of Michigan catalog number 11156, a skeleton originally assigned to *Ostodolepis* by Case (1929). Carroll and Gaskill (reluctantly) based their separation on differences in the development of the intercentra, definition of the neurocentral sutures, and the shapes of the neural spines. The description of its vertebral column here depends primarily on the accounts of Case (1929) and Carroll and Gaskill (1978), and the reader is directed to illustrations from those references.

Atlas-axis complex. Both Case (1929), and Carroll and Gaskill (1978) designated separate atlantal and axial vertebrae in *Pelodosotis*. As in *Pantylus* and other microsaurs, the anterior portion of the atlas is wider than that posterior for articulation with the basioccipital and exoccipitals. Unlike *Pantylus*, there is no distinct "odontoid" process. A pair of unfused atlantal neural arches are present, bearing normally oriented posterior zygapophyses, but no apparent articulations for proatlantal structures. The atlantal arches do not support neural spines.

The axis is only slightly different from the following vertebrae. It is longer in anteroposterior measure, with a correspondingly longer, but lower, neural spine. The spine is not developed like those of other microsaurs or primitive reptiles, and apparently did not provide a strong anchor for anterior epaxial musculature.

Dorsal vertebrae. Carroll and Gaskill (1978) reported an uninterrupted series of 44 dorsal vertebrae as well as a short string of three, the second of which was interpreted as a sacral. Minimally, 45 presacrals were present, and Carroll and Gaskill point out that there may have been at least one more. In either case, *Pelodosotis* has the largest presacral count of any known microsaur.

The centra are spool-shaped in a manner typical of microsaurs, and their anteroposterior length exceeds their height. Intercentra have not been found in articulation with the pleurocentra of the long articulated string, but elements identifiable as such are present in the specimen.

The zygapophyseal planes are steeply tilted, at angles ranging from 20 to 30 degrees. The transverse processes reach well beyond the lateral limits of the zygapophyses. Their costal articulations appear almost dumbbell-shaped, reminiscent of the condition in most captorhinids, but the parapophyseal articulations reach beyond the anterior edge of the centrum to the intercentrum. The intercentra are notched posterodorsally for reception of the capitulum of the ribs.

The neural arches in *Pelodosotis* are fused to the centra, though the lines of sutural attachment remain clear. The arches are not swollen, but are more heavily built than those in *Ostodolepis*. The neural spines alternate in height throughout the column. Both the tall- and low-type neural spines of *Pelodosotis* are more squared off in lateral view than those in *Ostodolepis* or *Micraroter* (Figure 24B; Carroll and Gaskill, 1978: Fig. 117H and I). The anterior border of tall-type neural spines rises almost vertically, as opposed to the less steeply angled neural spines of *Ostodolepis*. Similarly, the anterior and posterior margins of the low-type spines in *Pelodosotis* rise almost vertically, then flatten off to a horizontal ridge. The relative difference in height between tall- and low-type spines is comparable to that in *Ostodolepis*, but tall- and low-type spines in *Pelodosotis* do not appear to differ significantly in transverse width.

Except for the first two dorsals, the pattern of alternation proceeds uninterrupted throughout the length of the presacral column in *Pelodosotis*. The height of the third neural spine is greater than that of the fourth, but posteriorly all other odd-numbered spines are low and all even-numbered spines are tall.

Sacral structures. Both Case (1929) and Carroll and Gaskill (1978) identified a single broad-arched vertebra near the end of the preserved column as a sacral. Its identity was based on its differing arch construction, though it must be noted that the sacral structures of other Paleozoic tetrapods, including *Micraroter*, are usually indicated by a narrower, rather than wider, neural arch. Carroll and Gaskill speculated that it may be a presacral, as at least one specimen of another microsaur, *Micraroter*, has wider neural arches in the posterior portion of its presacral column.

The sacral vertebra has a low neural spine, with laterally placed, anteroposteriorly directed furrows. The configuration is very much like that of the low-spined vertebrae of many captorhinid reptiles. The shape of the centrum of the sacral vertebra of *Pelodosotis* is similar to those of the preceeding vertebrae.

Micraroter

Daly (1973) based the initial description of *Micraroter erythrogeios* on FMNH UR2311, a specimen that contained an almost complete skull but portions of only four associated vertebrae. Carroll and Gaskill (1978) later attributed a specimen housed at the Bernard Price Institute for Paleontological Research (BPI) Johannesburg, South Africa to the genus. Although Carroll and Gaskill (1978) placed both specimens in the same genus, they noted a number of differences, most notable the apparently juvenile condition of Daly's specimen. Only the Field Museum specimen was available for study, and little overlap exists in the represented vertebral materials of the two specimens. Description of the anterior portion of the vertebral column is based primarily on the holotype, whereas description of the more posterior regions of the column here depends on that by Carroll and Gaskill (1978).

Tentative calculations of the presacral length can be based only on the BPI specimen. Carroll and Gaskill (1978) estimated that 32 presacrals were present. A short gap behind the 25th presacral was probably occupied by a minimum of two additional presacrals. The presacral column could have been longer, but there is no reason to doubt this estimate.

Atlas-axis complex. The first two presacral vertebrae are the only two completely preserved elements of the vertebral column in FMNH UR2311. Their constituent parts are not fused, indicative of its immaturity. The body of the atlantal centrum is trapezoidal in ventral view (Daly, 1973: Fig. 23). A pair of anterolaterally directed processes protrude from the anteroventral surface of the centrum. Daly interpreted them as points of articulation with the exoccipitals. Carroll and Gaskill (1978) speculated that they are actually anteriorly directed transverse processes, and that the exoccipitals may have provided partial support for the atlantal rib (though they admit that such an arrangement seems unlikely). The exoccipitals are not well preserved in the holotype, but their posteromedial orientation and the low position of the processes seem to provide stronger support for Daly's argument.

Anteriorly, the atlas bears a strap-shaped articular face similar to that of other microsaurs, but an "odontoid" process is not highly developed in FMNH UR2311. The posterior third of the centrum narrows behind the anterolaterally directed facets. Parapophyseal and diapophyseal facets are located low on the lateral surface of the centrum.

The pedicels of the atlantal neural arch are only about one-third the length of the centrum. They fuse neither medially nor to the atlantal centrum in FMNH UR2311, but do so in the more mature BPI specimen. The atlantal arches do not support median neural spines, but each arch bears a posteriorly directed process. Daly (1973) hypothesized that ligamentous structures may have continued from them posteriorly, a condition analogous to the paired Mm. interspinales that run lateral to the low-type neural spines of captorhinid reptiles.

The second presacral ("axis") in *Micraroter* exhibits a more conventionally shaped centrum. It is pinched laterally (Daly, 1973: Fig. 21; Carroll and Gaskill, 1978: Fig. 115F), producing a slight ventral keel. A transverse process is easily distinguished on the lateral aspect of the second vertebra; however, a parapophyseal articulation is not obvious. The anterior zygapophyses extend far forward for support of the posterior atlantal zygapophyses, and are tilted at angles of 30 to 40 degrees. The spine is short and rounded in cross-section in FMNH UR2311, and is transversely narrow and more acuminate than those of the following vertebrae. Although not designated as an axis by Carroll and Gaskill (1978), the second presacral in the BPI specimen has a conspicuously developed neural spine, taller than those of any other presacral. Although not as long in anteroposterior measure, it angles anteroventrally in a manner similar to that in *Captorhinus* (Carroll and Gaskill, 1978: Fig. 60A). An intercentrum separates the atlantal centrum from the subsequent pleurocentrum.

Dorsal vertebrae. The dorsal vertebrae in the BPI specimen are very similar to those of *Pelodosotis*, though the former has a smaller presacral count and a greater degree of regional variation. The transverse processes are long anteriorly, shortening toward the rear of the column. Conversely, the zygapophyseal width increases posteriorly. Intercentra are present throughout the length of the presacral column; they are shorter in dorsal height but longer in midventral length than those in *Pelodosotis*.

The neural arches are low, but not swollen, describing a concave outline in anterior view. The neural spines alternate in height through much of the column. The 3rd to 5th spines are all shorter than the 2nd, but the 6th and 7th are considerably taller. An irregular alternation continues back to the 15th presacral, resumes from numbers 20 to 25, and again from the 29th through the sacral neural spines. The lateral outlines of both tall- and low-type spines in the BPI specimen approach that in *Ostodolepis*.

Sacral vertebrae and ribs. The BPI specimen of *Micraroter* is unique among microsaurs in possessing three sacral vertebrae; all others have one or two. As in most labyrinthodont amphibians and primitive amniotes, the anterior zygapophyses of the first sacral are equal in width to those of the presacrals, whereas the zygapophyseal widths of the other sacrals are markedly narrower. The centra are spool-shaped, but the presence of intercentra cannot be confirmed. As expected, the transverse processes are stoutly developed.

The BPI specimen is the only specimen known in which the sacral vertebrae display alternation of neural spine height. The first and third have low-type spines, the second a tall-type spine.

The three pairs of sacral ribs are equally developed. The tubercular and capitular heads are slightly separated. They display little change in thickness from the heads to their distal articulations. The first pair angles back to the anterior edge of the ilium, the second pair extends directly laterad, and the third pair angles forward. The distal edges of successive sacral ribs overlap slightly.

Caudal vertebrae and ribs. Other than *Pantylus*, the BPI specimen of *Micraroter* is one of the few microsaur specimens to include most of the tail. The presence of only 16 caudals suggests that the tail was short, probably coming to a stubby terminus. Central length remains almost constant, but both central and neural arch width decrease quickly. The transverse processes fade out around the 8th caudal. Alternation of neural spine height is not evident in the caudal series. Long haemal arches (about twice the length of the centra) are present from the 3rd caudal back, decreasing in length by about 75% through the end of the preserved caudal series. Caudal ribs are present back to the 14th caudal. The first 5 are the only ones that may be described as definitely bicipital. They are all sharply recurved, gradually shorten posteriorly, and come to a pointed termination.

Presacral ribs. Bicipital ribs are associated with all of the presacral vertebrae. They are cylindrical in cross-section, except for numbers 2 to 10, which are flattened distally. The 2nd to 5th are quite wide and strongly recurved. They extend parallel to the long axis of the trunk and presumably provided support for the pectoral girdle via serratus musculature. The ribs slowly increase in length to number 13. They do not decrease in size until just before the sacrum.

TRIHECATONTIDAE: *TRIHECATON HOWARDINUS*

Trihecaton howardinus is known from only the holotypic specimen, UCLA VP 1743, a nearly complete postcranium with associated dentary and maxillary fragments, and one referred specimen, UCLA VP 1744, which includes caudal vertebrae and scales from what is presumably the same individual.

The unique nature of *Trihecaton* prompted Vaughn (1972) to establish a new family, the Trihecatondidae. Trihecaton may be quite primitive in that it is the only microsaur known to possess labyrinthine infolding of the marginal teeth. Carroll and Gaskill (1978) speculated that it may be closely related to members of the family Ostodolepidae, all of which display alternation of neural spine height. Unfortunately, the lack of diagnostic cranial elements prevents detailed comparisons between the families. Of the postcranial materials available for study, only the configuration of the atlanto-axial vertebra is clearly similar between the two families.

Although the first vertebra is separated from the rest of the column, the holotype does not appear to be missing any presacrals. As far as can be determined, *Trihecaton* possessed 36 presacrals.

Atlanto-axial vertebra. The first vertebra in *Trihecaton* is very similar to that of the BPI specimen of *Micraroter.* Anteriorly, the "odontoid" process and articular surface conform to the general microsaurian pattern described by Carroll and Baird (1968) and Carroll and Gaskill (1978). Posteriorly, the atlanto-axial centrum is only half as wide as its anterior articular surface, and the notochordal pit is deep. Because of the marked difference in width from anterior to posterior, the sides of the atlanto-axial centrum face posterolaterally. The diapophyseal and parapophyseal articulations are rounded pits, clearly outlined by raised lips of bone about their circumference. The parapophyses are buttressed by a posteriorly directed ridge. Both costal facets are located low on the lateral surface of the centrum. No atlanto-axial ribs are present in the specimen, but they must have had distinctly separate and posteriorly directed tubercular and capitular heads.

The atlanto-axial neural arches are firmly fused to the centrum. They bow outward as they reach around the neural canal, almost touching at their anterodorsal limits; however, posteriorly they open out to delineate a wide neural canal. Stubby neural spines rest on each arch, diverging slightly from each other at their dorsal tips. They are not fused, leaving the neural canal dorsally exposed as in the ostodolepids *Pelodosotis* and *Micraroter.*

Dorsal vertebrae. The anteriormost dorsal vertebrae afford clear views of the ventral and lateral surfaces of the centra. They are hourglass-shaped and pinched laterally, but not so much as to create a ventral keel. Heavy lips mark the ends of the centra, which are bevelled ventrally for reception of the large intercentra (Figure 24C).

The intercentra (Figure 24D and E) are also slightly pinched laterally, extending farther posteriorly at the midline than the rest of the element. They are heavy and robust for a microsaur, and extend almost to the dorsal edge of the pleurocentra. They are also quite long anteroposteriorly. Their dorsal tips are embayed for reception of the capitulular head of the rib.

The transverse processes range from oval to kidney-shaped in outline, their articular faces directed ventrolaterally and just slightly anteriorly. The long axis of the costal facet is oriented anteroventrally toward the posterior embayment of the intercentrum. Just ventral to the transverse process, a gently curved suture marks the boundary of the centrum and neural arch. The pedicel of the neural arch is approximately two-thirds the length of the centrum.

Alternation of neural spine height and structure can be seen throughout most of the axial column in *Trihecaton.* The phenomenon is particularly evident in presacrals 17-21 and 26-35. As in other microsaurs, alternation of spine height does not appear to be restricted to regions near the limb girdles. Although absolute differences between tall- and low-type neural spines are not as great as in other forms that display alternation, they are nonetheless distinct. Low-type spines are low, transversely rounded ridges, that extend the entire midline length of the neural arch (Figure 24F). Tall-type spines are similarly constructed anteriorly, but their posterior half rises and usually diverges into two taller processes (Figure 24G). A few of the more anterior tall-type spines are undivided dorsally.

The neural arches appear almost heart-shaped in dorsal view (Figure 24F and G). None are swollen, but those associated with low-type spines are considerably heavier and more convex at their lateral limits than those supporting tall-type spines. Additionally, deep anteroposteriorly directed furrows run lateral to the low-type spines. These depressions cannot be seen in the lateral-view reconstructions of Carroll and Gaskill (1978). No such furrows are associated with the tall-type spines. The furrows are positioned directly in line with the paired processes of tall-type neural spines. Although the neural arches of *Trihecaton* are not swollen like those of captorhinid reptiles, it appears that they also allowed structures to run alongside the low-type spine through well-defined furrows. Alternation in *Trihecaton* is not limited to areas near the limb girdles, but the ridge-and-trough configuration is particularly well defined in the posteriormost dorsals, just anterior to the pelvic girdle.

Sacral structures. Only one sacral vertebra is recognizable. Because it is obscured by the last presacral, little more can be said about the sacral than to note its generally bulkier construction. The sacral neural arch appears to have been heavy and somewhat rounded in transverse section. The neural spine of the last presacral is tall, and the sacral neural spine also appears to have been stout. Alternation probably did not continue into the sacral region as in *Micraroter*. No sacral ribs are visible in UCLA VP 1743 or 1744.

Caudal vertebrae. UCLA VP 1744 includes 12 caudal vertebrae, all with associated haemal arches about twice the length of the caudal centra. A bridge of bone completes each haemal canal at its dorsal limit. Although relatively narrower than the presacral centra, the caudal centra also have rounded lips at each end and are slightly pinched laterally. The neural arches are narrow and firmly ankylosed to the centra; no suture lines are apparent. The caudal neural spines are triangular and pointed and, unlike *Pantylus*, there is no evidence of alternation in height in the tail region. Transverse processes are present on the anteriormost caudals in UCLA VP 1744, but quickly decrease in size posteriorly. They are, however, present as narrow, stubby ridges throughout the remainder of the preserved caudal series, and small ribs were probably present at least as far back. No caudal ribs are preserved in either the holotype or referred specimens

Dorsal ribs. Ribs are present throughout the presacral column. The anteriormost ribs are not well preserved, and the presence of distally dilated ribs near the shoulder girdle cannot be confirmed. The tuberculum in *Trihecaton* is flat, but the capitulum is distinctly rounded. Unlike the other microsaurs described thus far, the costal heads are joined by a thin web of bone. The thin web is continued distally as a shallow depression that runs along the proximal fifth of the rib. The ribs extend laterally, with little posterior curvature.

MOLGOPHIDAE

Molgophis macrurus and *Pleuroptyx clavatus*, of the "lepospondyl" family Molgophidae, are among the most enigmatic of the Paleozoic tetrapods treated in this study, and are included in this section on the Microsauria for the sake of convenience. Both are long, snake-like forms from the Middle Pennsylvanian Diamond Coal Mine of Linton, Ohio (approximately equivalent in age to the Westphalian D of the European Upper Carboniferous).

Both genera were originally described by Cope (1868, 1875) on the basis of partial vertebral columns. Romer (1966) considered *Pleuroptyx* to be a junior synonym of *Molgophis*, and placed the Molgophidae within the order Microsauria. Carroll and Gaskill (1978) interpreted the Molgophidae and Lysorophidae as closely related groups, distinct from microsaurs. Hook and Baird (1986) provisionally placed both genera in the Microsauria (sensu lato) and have acknowledged the possibility that *Pleuroptyx* may be a junior synonym of *Molgophis*. In an earlier study, however, Hook retained *Pleuroptyx* as a valid genus (Wellstead; in Hook, 1983). Wellstead (1985) has revised the taxonomy of the Lysorophidae and Molgophidae. Although Wellstead's conclusions may require taxonomic revisions concerning the specimens described, the morphological and functional interpretations described here should remain valid for these (closely related) forms.

Pleuroptyx clavatus

Only vertebral material may be confidently assigned to *Pleuroptyx*. A questionably associated hindlimb was ascribed to the genus by Cope (1875: Pl. XLIV, Fig. 2A), but Hook (1983) argued that the specimen (AMNH 6839) actually belongs to an indeterminate temnospondylous amphibian. AMNH 6863 consists of 12 presacral vertebrae. They are similar in most respects except for variability in the structure of the neural spines.

The centra (Figure 25C) are hourglass-shaped; and as they are quite closely articulated, the presence of intervening intercentra seems improbable. The neural arch pedicels are fused to the center of the centrum and measure only about half its length. Although ribs are clearly present, transverse processes are difficult to distinguish. Anterior and posterior zygapophyses reach to the ends of their associated centrum.

The neural arches are short and narrow. Most of the neural spines are of only moderate height (Figure 25C). In lateral view the spines are squared off, with the anterior border somewhat more oblique than the posterior. Two of the neural spines, numbers 9 and 11 of the string, are even shorter than those of the rest of the series. The low-type spines are only rounded nubbins and lightly constructed. Due to the compression of Linton materials, it is difficult to determine the transverse width of the low spines of AMNH 6863, but they do appear to be somewhat thinner than the others. Two low-spined vertebrae in a series of 12 do not allow for an interpretation of the pattern of alternation. However, they are sufficient for tentative confirmation of the presence of alternation in the dorsal series of this specimen.

Cope (1875) distinguished *Pleuroptyx* from *Molgophis* by the unusual construction of the ribs. Those of *Pleuroptyx* (Figure 25D) are short and curved, distally spatulate, and have wing-like processes on their posterior margins. As serial differences in costal structure are not uncommon among Paleozoic tetrapods, such may not be sufficient to distinguish *Pleuroptyx* and *Molgophis*, but the posterior alar processes remain problematic.

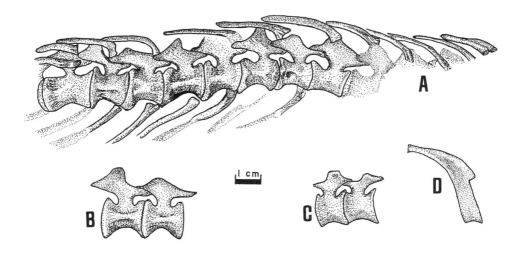

Figure 25. Axial structures of molgophid amphibians. A, MCZ 3326, latex peel of AMNH 6840, *Molgophis macrurus*; right lateral aspect of 10 dorsal vertebrae and associated ribs. B, reconstruction of dorsal vertebrae of *Molgophis macrurus*, right lateral aspect. Based on MCZ 3326. C, reconstruction of dorsal vertebrae of *Pleuroptyx clavatus*; left lateral aspect. Based on MCZ 3327. D, left dorsal rib of *P. clavatus*, left dorsolateral aspect. Based on MCZ 3327.

Molgophis macrurus

Wyman (1858: Fig. 2) provided the first illustration of the variability in neural spine height now recognized as the phenomenon of alternation. The specimen was at the time left unnamed, and Cope (1868) later named it *Molgophis macrurus*. Hook (1983) and Hook and Baird (1986) have argued that *M. macrurus* is the only valid specific designation for the Linton forms assigned to various species of the genus *Molgophis* (Cope, 1875).

The centra in *Molgophis* (Figure 25A and B), visible in AMNH 6840, are similar to those in *Pleuroptyx*. They are deeply spool-shaped, with lips developed at their anterior and posterior ends, the posterior lip slightly larger. Transverse width is difficult to determine, but the centra were probably rounded in end view. Ridges mark the lateral and ventral aspects of the centra. No intercentra are present in the short string available for inspection. Delicate anterior zygapophyses extend to the level of the anterior end of the centrum. The articular planes of the anterior and posterior zygapophyses are tilted, but compression of the specimen precludes determination of the exact degree of inclination. The neural arches in *Molgophis* are lightly constructed and somewhat narrow.

The neural spines are narrow and almost blade-like. Cope (1875) described the neural spines of *Molgophis* as "imperfectly preserved." In fact alternation in neural spine height is clearly exhibited through most of AMNH 6840. The first three spines are tall. They do not reach as far forward as the anterior limit of the neural arch, but angle steeply from its midpoint to the spine apex and then downward to the posterior margin of the arch. Thereafter, low spines alternate with tall spines to the end of the series. The anterior and posterior faces of the low-type neural spines slope up to the apex more gradually than in the taller types, resulting in a height only about one-third to one-half that of adjacent taller spines.

Transverse processes are not clearly evident, but well-developed ribs are present through the length of the specimen. A confident description of the structure of the costal heads cannot be made, but they appear to be holocephalous. The shafts are not strongly recurved. Although Cope (1875) distinguished *Molgophis* from *Pleuroptyx* on the basis of costal structure, the possibility that the long-bodied molgophids may simply have exhibited regional variation in rib structure cannot be dismissed.

MICROSAURIA: SUMMARY

The patterns of regional occurrence of alternation in neural spine height found among microsaurs are not as clearly defined or restricted as those of captorhinid reptiles or diadectomorph amphibians. Microsaurs exhibit considerable variation in: (1) relative differences in height between adjacent neural spines of different regions of the vertebral column, (2) patterns of alternation among different taxa of the order, and (3) the structure of the neural spines that display the alternation.

Despite the differences, both among members of the Microsauria and between it and other groups of Paleozoic tetrapods, certain broad similarities remain. With the prominent exception of the molgophids, the microsaurs that exhibit alternation of neural spine height tend to be terrestrial forms, with well-developed limbs. Although their limbs are sometimes short relative to body length, they are usually complete and well ossified. None of the extremely small-limbed microbrachids are known to show alternation of neural spine height. *Pantylus* was certainly a fully terrestrial tetrapod; and whereas the ostodolepids and *Trihecaton* may not have been quite so fully adapted to life out of the water, their short but well-defined limbs were probably capable of terrestrial excursions of some length. *Trihecaton* does exhibit an emphasis of the alternation just anterior to the pelvic girdle, in a manner similar to the captorhinids, and *Pantylus* may have done so as well.

5

THE PELYCOSAURIA: *VARANOSAURUS*

Traditionally, the order Pelycosauria has been considered among the earliest known of primitive reptiles in the fossil record. More recently, Gauthier and his coworkers (Gauthier et al., 1988) have removed the Pelycosauria from the Reptilia, retaining it as the most primitive grouping of the Amniota. Romer and Price (1940) provided a monographic review of the pelycosaurs, and more recently, detailed comparisons have been made among the best known genera (Brinkman and Eberth, 1983; Reisz, 1980, 1986). However, certain genera remain poorly known, among them the Early Permian form *Varanosaurus*. It was assigned to the Ophiacodontidae, at the time considered to be the most primitive family of pelycosaurs, by Romer and Price (1940). However, the primitive position of ophiacodonts has since been questioned (Brinkman and Eberth, 1983; Reisz, 1980, 1986).

The vertebral structures of *Varanosaurus acutirostris* have recently been redescribed (Sumida, 1989b), and the taxonomic history of the genus is included therein. Although the following description and analysis is abridged in part from that work, it is repeated here in large part for the sake of completeness in this review of alternation of neural spine height and structure in Paleozoic tetrapods. Further, although *Varanosaurus* displays certain cranial features typical of ophiacodontid pelycosaurs, some of its most characterisitc defining features are those of the vertebral column. Specimens of significance to this study include: BSPHM 1901 XV20 (the type specimen), AMNH 4174, and FMNH PR1670.

Both Case (1917) and Romer and Price (1940) reported 27 presacral vertebrae (including the atlas and axis) and 2 sacral vertebrae in *Varanosaurus*. Examination of AMNH 4174 indicates that the 12th presacral is labeled 12 on the centrum and 13 on the neural arch. The next vertebra is labeled 14, effectively adding one to the presacral count. However, one isolated neural arch included with AMNH 4174 appears to belong to the specimen. As the segments of the preserved column are not continuous, the actual presacral count may be 27 or higher.

Atlas-axis complex. Structures of the atlas-axis complex are present in all specimens with reasonably intact cranial materials: AMNH 4174, BSPHM 1901 XV20, and FMNH PR1670. Although the atlas-axis complex bears a number of resemblances to that in *Ophiacodon*, Reisz (1986) has noted that unlike the condition in *Ophiacodon*, the atlantal centrum does not fuse with the axial intercentrum in *Varanosaurus*. This does appear to be the case in FMNH PR1670, which is probably a juvenile specimen. On the other hand, AMNH 4174 does show fusion between these two elements (Figure 26B). Further, the elements are not always fused in ophiacodonts (Romer and Price, 1940: 206). The degree of fusion between the two elements appears to be a function of maturity rather than one of taxonomic distinction. A notch on the anterior face of the composite element indicates a posteroventral plane of fusion, an angle opposite that in *Ophiacodon* (Romer and Price, 1940: Fig. 44A and C). The centrum is disc-shaped. Articular facets for the atlantal neural arches are directed anterolaterally. The fused axial intercentrum precludes exposure of the atlantal centrum on the ventral surface of the column and possesses small parapophyseal articulations for the axial ribs, which are thinner and narrower than those in *Ophiacodon*.

The atlantal intercentrum (Figure 26B) is a blocky element, similar in proportion to that in *Ophiacodon* and *Labidosaurus* (Sumida, 1987). Parapophyseal articulations for the axial ribs project ventrolaterally from either side.

The atlantal neural arch (Figure 26A and B) is a paired structure that did not fuse dorsally. Concave ventromedial facets of the neural plates articulate with either side of the anterior edge of the atlantal centrum. A slight constriction sets them off from the more dorsal atlantal neural arch proper. A posteriorly directed wing-like process resembles those in *Labidosaurus* (Sumida, 1987) and *Petrolacosaurus* (Reisz, 1981), though the posteriormost portions are somewhat more rounded. Posteriorly directed spines of the neural arch halves are not preserved in AMNH 4174, though breakage marks indicate that they were present and directed posteriorly. They do not appear to have been as large as those in *Ophiacodon*. No dorsally directed neural spine is present. Anteriorly, a narrow, horizontal ridge provided a point of articulation for the proatlas (Figure 26A).

The posterior half of the left proatlas is preserved in AMNH 4174; it arches slightly anterodorsally from its articulation with the atlantal arch (Figure 26B). Though not preserved, the anterior half of the proatlas appears to have lain close to the horizontal plane.

The axial centrum (Figure 26B) has a shallow, but well-defined, ventral keel. The posterior face is ventrally embayed for the intercentrum of the third presacral vertebra, but the anterior face is nearly vertical. Near the dorsal limit of the centrum, the long axis of the teardrop-shaped transverse process is horizontally oriented. A horizontally directed ridge extends posteriorly from the base of the transverse process back to the rim of the centrum. The neural arch pedicel is defined by a posterior emargination just above the level of the transverse processes. The articular planes of the zygapophyses are tilted at an angle of about 30 degrees. The neural arch is sturdy, but not "swollen." Most of the axial neural spine in AMNH 4174 is broken off, but is well preserved in FMNH PR1670. What remains of the former extends far forward, between the atlantal arches, up to a transverse plane intersecting the proatlantal articulations of the atlantal arches. In FMNH PR1670, the axial neural spine is an anteroposteriorly elongate blade. As in *Ophiacodon*, the spine is sharp at its anterior edge, but thicker posteriorly.

Dorsal vertebrae. The dorsal centra are amphicoelous. Although they lack a sharp ventral keel, they are slightly pinched laterally, resulting in bluntly triangular cross-

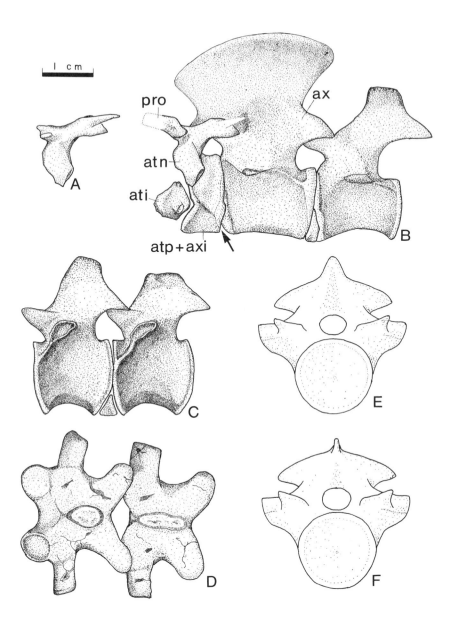

Figure 26. *Varanosaurus acutirostris*. A, reconstruction of left atlantal neural arch, left lateral aspect. B, reconstruction of atlas-axis complex and 3rd presacral, left lateral aspect. C, reconstruction of tall- and low-spined vertebrae in articulation, left lateral aspect. D, AMNH 4174, dorsal aspect of presacrals 7 and 8 showing difference in breakage patterns of tall- and low-type neural spines. E, reconstruction of tall-spined vertebra, anterior aspect. F, reconstruction of low-spined vertebra, anterior aspect. A and B based on AMNH 4174, BSPHM 1901 XV20, and FMNH PR1670. C based on AMNH 4174. Reconstructions in E and F are purposely simplified because of the inaccessibility of anterior and posterior aspects of vertebral materials in AMNH 4174.

sections (Romer and Price, 1940: Fig. 17). The anterior and posterior ends are only slightly recessed for reception of small intercentra.

The anterior zygapophyses are stoutly buttressed throughout the column. Although generally described as horizontal in *Varanosaurus*, toward the anterior end of the column the zygapophyseal facets are tilted at angles approaching 20 degrees. Those of the anterior pairs are slightly concave, whereas the posterior pair are correspondingly convex. The zygapophyseal facets of the middle and posterior dorsal vertebrae are tilted at an angle of about 15 degrees. The posterior vertebrae appear to be more tightly articulated at both the zygapophyseal and central articulations than those more anterior.

The transverse processes in *Varanosaurus* are well-developed, longest on the 6th and 7th presacrals where their length is close to three-fourths the centrum width. The transverse processes show a thin bony connection to the anterior edge of the centrum, a condition conspicuously like that in *Ophiacodon* (Romer and Price, 1940).

The neural arches of the axis and the 3rd and 4th presacrals following are similarly proportioned. The next two are wider; the 7th is distinctively swollen. The 8th neural arch and those following are swollen, with no evidence of "pits" or excavations. *Dimetrodon* and a number of other pelycosaurian genera, generally sphenacodonts, exhibit lateral excavations of the neural arch that sweep from the anterior zygapophyses to the middle of the neural arch. It has been presumed that these excavations lightened the vertebrae while retaining the strength of an arched construction. Other pelycosaurs also lack this excavation, but only *Varanosaurus* has swollen neural arches as well. Excavated neural arches and swollen neural arches were apparently independent solutions to the problem of strengthening vertebral support of the body.

Only two neural spines are preserved in BSPHM 1901 XV20, and their preservation varies in AMNH 4174. Examination of the preserved spines and the breakage marks of the missing spines in AMNH 4174 indicates that in some regions of the column neural spines alternated in height and structure (Figure 26C and D), and certain neural spines are distinctively taller than adjacent ones. These taller types are trapezoidal in lateral view and angle slightly posteriorly. "Low-type" spines are slender from side to side and more expanded anteroposteriorly, giving them the look of a narrow wedge (Figure 26F). The break marks also indicate two distinctive types of spines: the bases of the taller types are oval in outline, whereas those of the lower types are long, narrow, and rectangular (Figure 26D). Thus, a rough pattern of spine height and shape can be established for *Varanosaurus*. Presacrals 5, 8, 12, 14, 16, and 19 have low-type spines. Except for those vertebrae in which missing neural arches and spines preclude examination, all other spines appear to be tall types. A distinct pattern of alternation is clear only in the midcolumn region; it appears to be less regular in other areas.

Sacral vertebrae and ribs. As in primitive reptiles and ophiacodonts, two sacral vertebrae are present in *Varanosaurus* (Figure 27A). Both sacral centra are slightly shorter in length than those of the presacral series, but are more robustly built. The intercentrum of the second sacral is wedged tightly between the two vertebrae. The ventral surface of the second sacral centrum is slightly keeled.

The anterior zygapophyses of the sacral vertebrae are heavily buttressed, those of the first sacral even more so than the second. Immediately ventral to the anterior zygapophyses, the transverse processes are impressively proportioned. Those of the first sacral measure half the central length, and stout supports sweep forward from the posteriormost region of the centrum. Although not quite as long in anteroposterior measure, the transverse processes of the second sacral are also braced by a substantial bony base. A horizontal ridge extends from the transverse process of the second sacral back to the posterior edge of the centrum. The neural arch pedicels are defined by a posterior embayment above the centrum, but no comparable constriction can be seen anteriorly.

The generally heavy construction of the sacral vertebrae results in the anterior portions of their neural arches approaching the proportions of the presacral neural arches. The neural arches sweep smoothly up to sharp, narrow neural spines that do not contribute to any pattern of alternation in height. Both extend the entire anteroposterior length of the neural arch, and angle back slightly. A ridge runs parallel, and a short distance posterior, to the anterior border of the second sacral spine.

The first sacral rib (Figure 27A) is constructed quite robustly. Its more proximal portion extends laterally for a short distance before angling ventrolaterally. More distally, the rib expands into an extremely broad, flared plate which is primarily horizontal in orientation; however the anterior edge of the rib turns sharply dorsally. A series of fine, parallel ridges runs to the edge of its iliac surface. A thick, posteromedially oriented border meets the anterior articular surface of the second sacral rib.

The distal portion of the second sacral rib is not as broadly spatulate as the first, but is actually thicker in dorsoventral dimension. Its limited expansion is directed anteriorly to articulate with the first sacral rib. It has a short articulation with the ilium that measures only about one-sixth the length of the iliac contact of the first sacral rib. As in the first, the distal expansion of the second sacral rib lies close to the horizontal plane.

Caudal vertebrae and ribs. Case (1910) reported 32 caudal vertebrae in AMNH 4174. The specimen includes all or part of 32 caudals, but they are not a continuous series; a number of caudal vertebrae must be missing from the specimen. Three caudals are attached behind the second sacral vertebra, and there is a gap of what was probably at least five caudals between this and the next preserved string. *Varanosaurus* probably had a tail containing at least 45 to 50 vertebrae.

The heads of caudal ribs are preserved in association with only the most proximal caudal vertebrae. They appear to have had capitular and tubercular heads connected by an extremely thin web of bone. What can be seen of the costal shafts indicates that they were probably recurved.

The first three caudals show little decrease in size or change in proportion. In view of the series of vertebrae that is missing from the specimen, *Varanosaurus* may have had a relatively thick, heavy tail. Although stoutly built, the proximal caudal vertebrae did not have expanded neural arches. The spines of the first three caudals are broken off, but appear to have been transversely narrow. There is no evidence of alternation in height of caudal neural spines.

The centra of the proximal caudal vertebrae display a rounded midventral ridge. A thin web of bone connects the transverse processes to the anterior edge of the centrum. The zygapophyses tilt at an angle of approximately 25 degrees.

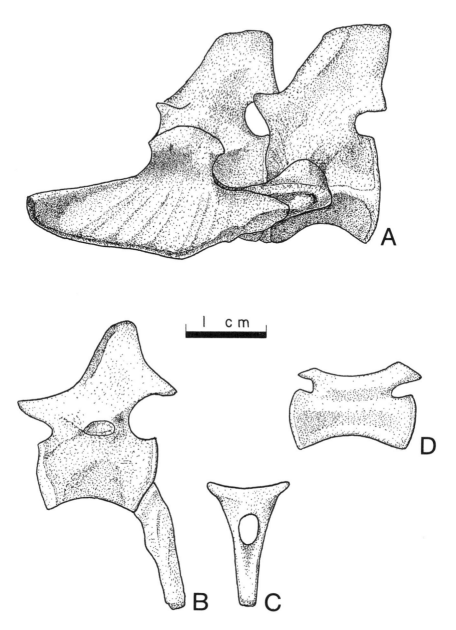

Figure 27. AMNH 4174, *Varanosaurus acutirostris*. A, reconstruction of sacral vertebrae and sacral ribs, left lateral aspect. B, anterior caudal vertebra and haemal arch, left lateral aspect. C, reconstruction of haemal arch, anterior aspect. D, mid-caudal vertebra, left lateral aspect.

Although central length does not change appreciably by the middle of the tail, the transverse measure decreases substantially. By approximately the 8th caudal, haemal arches are clearly present (Figure 27B and C), though they may have been present more anteriorly. In this region of the column the transverse processes are circular in outline, are located more dorsally on the neural arch pedicel, and do not retain a web-like connection with the anterior edge of the centrum. The caudal ribs must have been single-headed. The pedicels are more clearly defined than those of more proximal caudals. By the middle of the tail, the centra are narrower and strongly pinched laterally. The neural arches are also less stoutly built, with shallow excavations at their bases. These modifications of the centra and neural arches are somewhat similar to those seen in the dorsal vertebrae in *Sphenacodon* and *Dimetrodon*, but not as deep.

The centra of the posteriormost portion of the tail retain approximately the same length as those of the more proximal caudals. The lateral excavations of the centra define a distinct horizontal ridge. Haemal arches cannot be detected beyond the 30th caudal. Although the neural arches become very low and narrow, and the neural spines almost inconsequential in proportion, both persist in even the most distal elements of the preserved tail. The zygapophyses are tilted at 30-35 degrees. Transverse processes cannot be seen in the far posterior region of the tail.

Presacral ribs. Ribs were present throughout the presacral series. The few well-preserved transverse processes in AMNH 4174 indicate that capitular and tubercular heads were connected by a thin web of bone. AMNH 4174 gives little indication of the shape of the costal shafts, but Watson (1914) reported that ribs located near the pectoral girdle in the holotype bear distally spatulate ribs, like those of many captorhinid reptiles.

Intercentra. Except for the axial intercentrum, which is fused to the atlantal centrum, intercentra are present as distinct elements throughout the presacral series. They are typically wedge-shaped, but are quite small when compared to those of other forms that exhibit alternation of neural spine height. The sacral intercentra are also relatively small.

VARANOSAURUS: SUMMARY

As *Varanosaurus* is the only pelycosaur known to display both alternation of neural spine height and swollen neural arches, the possibility that they are strongly correlated must be considered. However, the wide diversity of groups in which alternation occurs points to the possibility that these features were derived in parallel. This suggests that selection for the functions provided by alternation in neural spine height was strong.

In *Varanosaurus*, alternation does not appear to be as clearly concentrated near the limb girdles as in captorhinids, some seymouriamorphs, and diadectomorphs. Alternation is extended for more than a few segments only in the mid-dorsal region, but the low neural spines of presacrals 5 and 8 may have extended its biomechanical effects to the region of the pectoral girdle.

Although the general pattern of alternation in *Varanosaurus* does not correspond to the patterns in other taxa displaying this phenomenon, the common possession of such indicates potential similarities in function. The patterns of epaxial musculature in regions of alternation were probably similar to those discerned in *Captorhinus* and *Labidosaurus*. Significant vertebral dorsiflexion is precluded in most pelycosaurs because of the presence of relatively tall, robust neural spines. "Low-type" spines alternating with "tall-type" spines of only moderate length in *Varanosaurus* would have reduced the potential for contact between neural spines during flexion. The combination of these factors would have allowed a greater potential for vertebral dorsiflexion in *Varanosaurus* than in other types of tall-spined pelycosaurs. The tilt of the zygapophyseal articulations and their slightly curved facets would also have allowed such dorsiflexion, as well as lateral flexion and a small amount of rotation about the longitudinal axis of the vertebral column. *Varanosuarus* may have been more agile than previously believed, and if these effects were communicated to the region of the shoulder girdle, alternation may have affected its locomotory behavior.

The possession of expanded neural arches in diadectomorph and seymouriamorph amphibians is most likely the reason Romer and Price (1940) regarded *Varanosaurus* to be a primitive, relict member of the Ophiacodontidae. Despite the apparently primitive condition of the vertebral column, its uncertain taxonomic position and the removal of the Ophiacodontidae from a primitive position at the base of the Pelycosauria (Brinkman and Eberth, 1983, 1986; Reisz, 1980, 1986) renders the use of *Varanosaurus* as a model for primitive pelycosaurs inappropriate (Sumida, 1989b).

6

ARAEOSCELIDIA

The reptilian suborder Araeoscelidia includes a number of gracile, lizard-like forms (including the Late Pennsylvanian diapsid *Petrolacosaurus*) from the Late Paleozoic of North America, and in one case from Europe (Vaughn, 1955; Tatarinov, 1964; Reisz et al., 1984). General concepts of the Araeoscelidia have centered on the genus *Araeoscelis* (Vaughn, 1955; Reisz et al., 1984), and more recently on *Petrolacosaurus* as well (Peabody, 1952; Reisz, 1977, 1981).

The araeoscelids have been subjected to various phylogenetic interpretations (Vaughn, 1955; Romer, 1956, 1966; Reisz et al., 1984), and *Araeoscelis* ranks among the most thoroughly studied of Late Paleozoic tetrapods (Vaughn, 1955; Reisz et al., 1984). The family Araeoscelidae also includes *Zarcasaurus* (Brinkman et al., 1984), and possibly two lesser-known genera: *Dictybolos* (Olson, 1970) and *Kadaliosaurus* (Credner, 1889). Of these three, only *Zarcasaurus* includes vertebral material of sufficient quality and preservation to warrant description in this study. *Petrolacosaurus kansensis* is the only member of the family Petrolacosauridae.

Vaughn (1955) and Reisz et al. (1984) have pointed out the intermediate position of the Araeoscelidia between primitive captorhinomorph reptiles and more advanced diapsids. Vaughn (1955), Reisz (1981), and Reisz et al. (1984), have provided thorough descriptions of the axial skeleton in *Araeoscelis* and *Petrolacosaurus*, but the importance of the Araeoscelidia as a transitional group, and the presence of alternation of neural spine structure in its best-known genera, justifies a brief review of their vertebral structure. Inasmuch as the above authors have provided extensive illustrations in their studies, their efforts are not duplicated here. Representative illustrations are provided for *Araeoscelis* and *Petrolacosaurus*; additionally, the reader is referred to the appropriate figures of Vaughn (1955) and Reisz (1981).

ARAEOSCELIS

Williston (1910) first described *Araeoscelis gracilis* and included illustrations and descriptions of much of the axial skeleton. Broom (1913) later described the closely related *Ophiodeirus casei*. As part of an exhaustive study of the genus *Araeoscelis*, Vaughn (1955) recognized the extreme similarity between the two taxa and declared *Ophiodeirus* the junior synonym, but retained the species name on the basis of stratigraphic separation. More recently, Reisz et al. (1984) have provided a description of two well-preserved partial specimens. As he could not distinguish between the two species on the basis of morphological criteria, Vaughn treated all of the *Araeoscelis* materials available to him as a single form. The same convention is followed here.

Vaughn (1955) estimated the presacral count in *Araeoscelis* to be 31. Based on more complete specimens, Reisz et al. (1984) estimated that the preascral count is more likely 28 or 29. As the cervical and dorsal regions of araeoscelids display distinct structural differences, they will be described separately.

Atlas-axis complex. The presence of proatlantal elements has not been confirmed to date, though facets for their articulation with the braincase project from the exoccipitals (Vaughn, 1955). As in many other Late Paleozoic tetrapods, the atlantal centrum and axial intercentrum are fused (Vaughn, 1955: Fig. 5A and C-E). The ventral aspect of this composite element is slightly pinched laterally, resulting in a rounded midventral ridge. The atlantal centrum is considerably shorter than any of the following cervical vertebrae. No costal facets can be seen. The anterior face of the atlantal centrum in *Araeoscelis* is pentagonal in shape, with a dorsal horizontal border. Most of the face is flat, except for a small, anteriorly produced process with a concave pit for articulation with the basioccipital. The atlantal intercentrum in *Araeoscelis* is poorly known; as indicated by the anterior depression of the atlantal centrum, it was apparently a crescent-shaped element.

Only fragments remain of what may have been the atlantal neural arch halves in *Araeoscelis*. Vaughn (1955) indicated, however, that each half did not fuse medially, and that they probably had a posteriorly directed spine, extending beyond the plane of articulation between the atlantal and axial centra.

The axial centrum (Vaughn, 1955: Fig. 5A) is a long, amphicoelous cylinder, approximately three times the length of the atlantal centrum. Its midventral, longitudinal ridge is somewhat sharper than that of the atlantal centrum. Toward the anterior end of its lateral surface a blunt nubbin may indicate the position of a costal facet, but its exact structure is not clear. The pedicels of the neural arch are proportionately elongate. The zygapophyses are closely approximated, and the neural arch shows no indication of lateral expansion. The anterior zygapophyses are unusual in that the articular facets face dorsolaterally. The axial neural spine is larger than in any other vertebrae in *Araeoscelis*; it is greatly elongated anteroposteriorly. A rounded anterior projection reaches forward above the atlantal centrum. The anterior dorsal border of the spine is nearly horizontal, but posteriorly it gradually slopes ventrally to the posterior zygapophyses. The lateral surfaces of the axial neural arch are concave.

Cervical vertebrae. Vaughn (1955) proposed two criteria for identification of cervical vertebrae in *Araeoscelis:* elongate centra and the presence of a single costal facet not divided into capitular and tubercular articulations. Vaughn estimated that *Araeoscelis* possessed nine cervicals (including the atlas and axis), a number confirmed by Reisz et al. (1984) with more complete specimens.

As mentioned above, the centra are elongate. Their anteroposterior length increases until the 4th, then decreases. A midventral ridge runs the length of all the cervical centra. Anteriorly the transverse processes are short, ventrolaterally directed nubbins, but in the posteriormost cervicals they project more strongly laterad, and begin to show the development of distinct tubercular and capitular articulations. The transverse process is positioned at the anterior end of a horizontal ridge that runs along the lateral surface of the centrum. The posterior surfaces of the centra are inclined approximately 10 degrees from the vertical.

Cervical neural arch pedicels are also long, approximately 80% the central length. The anterior and posterior zygapophyses are closely approximated, with an inward tilt of about 10 degrees. The neural arches are narrow and anteroposteriorly extended. By the end of the cervical series their posterior aspects are more transversely expanded, with outlines similar to those of the dorsal vertebrae. Lateral excavations of the neural arches are most strongly expressed in the 8th and 9th cervicals.

The neural spine of the 3rd cervical is triangular in lateral view, but those more posterior are long, low ridges, approximately as wide as they are high. After the 4th, their anteroposterior length decreases in proportion to other vertebral dimensions. Most of the cervical spines are inclined anteriad, but they gradually assume a horizontal orientation by the 9th. Mammillary processes project anterodorsally from the 6th to 9th cervical spines. Those of the 6th to 8th are situated near the anterior edge of the spine, whereas those of the 9th are more centrally located. The mammillary processes in *Araeoscelis* are unlike those in *Eocaptorhinus*, in which they are more clearly associated with the neural arches (Dilkes and Reisz, 1986). The cervical neural spines in *Araeoscelis* exhibit no alternation in height or structure.

Dorsal vertebrae. Williston (1914) reported 19 or 20 dorsals in *Araeoscelis*. Vaughn (1955) arrived at a figure of 22 dorsals, but was limited by the disarticulated nature of the materials with which he was working. Reisz et al. (1984) agreed with Williston's estimate based on two specimens, neither of which included a complete dorsal column, but did provide enough overlap of elements to allow a reasonable estimate.

Dorsal centra (Figure 28B and C) are amphicoelous, bevelled ventrally for the intercentra. They do not exhibit any substantial change in length through the dorsal series. The parapophysis is a vertically oriented oval in cross-section that extends to the level of the intercentrum. As in the cervical vertebrae, the neural arch pedicels are about 70-80% the length of the centrum. At the anterior edge of each pedicel is an obliquely oriented diapophysis. The diapophyses and parapophyses become thin, minute ovals more posteriorly, the former disappearing about 4 or 5 vertebrae from the sacrum. The transverse processes extend well beyond the lateral edge of the neural arches in anterior dorsal vertebrae, but decrease in length posteriorly. The zygapophyses are more broadly spaced than those of the cervical series, the anterior pair providing a shallow concavity for the slightly convex facets of the posterior pair. The zygapophyseal facets are angled inward at about 20 degrees from the horizontal.

The neural arches of dorsal vertebrae in *Araeoscelis* are transversely widened more posteriorly in the column, and the lateral surfaces are marked by fossae-like structures. These two characteristics give them a bluntly X-shaped outline in dorsal view (see Vaughn, 1955: Fig. 6A and C). The edges of the lateral fossae may have been the site of attachment for M. spinalis dorsi and the M. semispinalis dorsi.

Romer (1956) interpreted the neural spines in *Araeoscelis* as being poorly developed. In fact, they are well developed, and manifest alternation in structure through the dorsal series. Alternation in neural spine construction has its anteriormost expression in the last cervical and first three dorsals (C9, D1, D2, and D3 of Vaughn, 1955). The neural spines of C9 and D2 are lengthened anteroposteriorly about 25-33% greater than those of C1 and C3. Beginning with D1, shorter anteroposterior length becomes associated with a taller, more acuminate spine, whereas the other spines retain the long, low form typical of low-type neural spines. Reisz et al. (1984) recognized alternation of neural spine height from presacral 17 posteriorward. MCZ 2304 (MCZ 4 and MCZ 7 of Vaughn, 1955) indicates that, in at least some specimens alternation was present more anteriorly as well. By combining the pattern displayed by MCZ 2304 with the reconstruction provided by Reisz et al. (1984), alternation of neural spine height would be suggested for the entire dorsal series, with odd-numbered presacrals as low types and even-numbered presacrals as tall types. The pattern terminates with a low neural spine on the last presacral, followed by a tall, well-developed spine on the first sacral. Short strings of vertebrae whose serial positions can be identified (MCZ 2043, MCZ 4383) appear to confirm the pattern described above. Mammillary processes positioned near the posterior border of the neural spine are present from the anteriormost dorsal to the mid-dorsal region.

Vaughn (1955) interpreted the position of the pectoral girdle as farther back than did Reisz et al. (1984). In Vaughn's reconstruction, alternation of neural spine height would begin at a level corresponding to the anterior edge of the pectoral girdle, approximately three vertebral segments anterior to the glenoid facet. The reconstruction proposed by Reisz et al. would place the initiation of the phenomenon directly dorsal to the glenoid. The pattern expressed among the anterior dorsals of certain specimens of *Araeoscelis* is not as strongly expressed as the alternation seen anterior to the pelvic girdle. The extent of alternation through the dorsal series would have allowed its associated movements to influence both the pectoral and pelvic limbs, but it probably had a more pronounced biomechanical effect on the latter in *Araeoscelis*.

*Sacral vertebrae and rib*s. There are two sacral vertebrae in *Araeoscelis*. The first is recognizable by its narrowing of the separation of the posterior zygapophyses. The zygapophyses of the second sacral are similarly approximated. The centra do not differ significantly in length or width from those of the dorsal series. The intercentrum of the second sacral is fused to the posterior face of the first sacral centrum.

Structures of the first sacral vertebra are largely determined by the large facets for the first sacral rib which covers nearly its entire lateral surface. The proximal articulation of the second sacral rib is as long as that of the first, but is not as thick dorsoventrally. Lateral excavations of the sacral neural arches are similar to those of the dorsal vertebrae, although their ventral extension is limited by virtue of the buttressing of the transverse processes. The neural spines are tall and more squared off than those of tall-spined dorsals. Both lean caudally, the second bearing a hooked posterior process in MCZ 2043 (Vaughn, 1955: Fig. 6H and I). Sacral neural spines do not contribute to a pattern of alternation.

The first sacral rib is more massive than the second, accounting for almost all of the iliosacral articulation. A shallow groove extends along the proximal edge of the first sacral rib. The large, spatulate, distal plate of the first rib angles posterodorsally in lateral view. The anterior portion of its iliac articulation is thicker and more rounded than its posterior portion or that of the second sacral rib. The second sacral rib angles

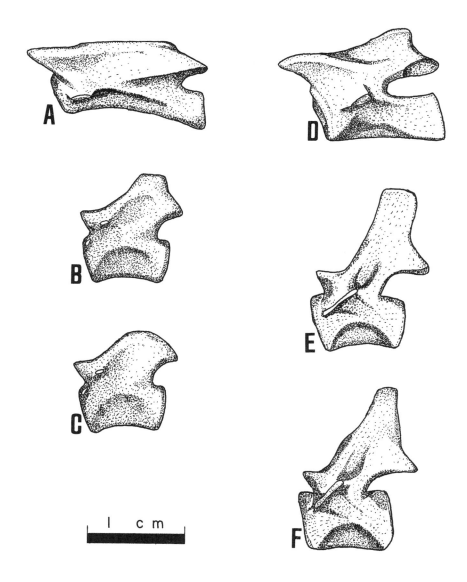

Figure 28. Left lateral aspects of *Araeoscelis gracilis* (A-C) and *Petrolacosaurus kansensis* (D-F) vertebrae. A and D, cervical vertebrae. B, tall-spined vertebra of *Araeoscelis*. C, low-spined vertebra of *Araeoscelis*. E, presacral 19, wide-spined vertebra of *Petrolacosaurus*. F, presacral 18, narrow-spined vertebra of *Petrolacosaurus*. A-C based on MCZ 2304 and USNM 22078. D-F based on KUVP 9951 and 9956.

forward to articulations with the first sacral rib and the ilium. Its distal terminus is considerably narrower than its proximal width.

Caudal vertebrae and ribs. Reisz et al. (1984) reported on the longest string of caudals known for *Araeoscelis* (17), from the most proximal portion of the tail. Caudal centra decrease gradually in length and width, but retain their ventral longitudinal ridge well into the tail. The generally lighter build of caudal vertebrae is reflected in their relatively shorter pedicels and closely approximated zygapophyses. The neural arches are narrow, and bear lateral excavations like those of the dorsal and sacral vertebrae. The excavations gradually disappear posteriorly in the tail.

The neural spines of the first two caudals in MCZ 2043 are angled strongly anteriorly, though other materials ascribed to *Araeoscelis* do not show this condition. All of the caudal neural spines are narrower than those of the dorsal series. Their apices are more rounded than those of the sacrals. Neural spines of the midcaudal region angle posteriorly and, although they decrease in height posteriorly, remain clearly discernible structures well back in the tail.

Transverse processes extend laterally from the 9 most proximal caudals, but only the first 4 bore true caudal ribs, which are strongly recurved. Discrete intercentra are present back to the 5th caudal. More posteriorly, intercentral outgrowths are present as haemal arches.

Intercentra. Intercentra exist as separate ossifications throughout the vertebral column in *Araeoscelis*, except for the axial intercentrum, which is fused to the atlantal centrum, and the second sacral intercentrum which is fused to the first sacral centrum. Intercentra are typically crescentic and are relatively smaller than those of captorhinid reptiles.

Presacral ribs. Cervical ribs are single-headed, with short, thin, tapered shafts. Apparently, those most anterior ran almost parallel to the longitudinal axis of the vertebral column (Vaughn, 1955). The 7th to 9th cervical ribs were recurved and set at a strongly acute angle to the longitudinal axis. At the apex of their curvature a short spine is directed forward (possibly for insertion of a M. levator costae; Vaughn, 1955). The 8th and 9th ribs are holocephalous, but display distinct capitular and tubercular areas.

Anterior dorsal ribs are clearly bicipital. More posteriorly, they become single-headed, with only a capitular articulation. Dorsal ribs are posterolaterally directed. The shafts are slightly flattened and distally dilated, but become shorter and thinner by the middle of the series. Vaughn (1955) noted a vertically oriented ridge just distal to the tubercular articulation. At approximately the same level a posterodorsally directed process marks the point of insertion of the M. iliocostalis. These processes fade rapidly toward the rear of the column.

ZARCASAURUS TANYDERUS

The holotype (CM 41704) and only known specimen of *Zarcasaurus tanyderus* is based on a small amount of disarticulated material that includes the complete or partial remains of five dorsal and three cervical vertebrae, including the axis. No atlantal, sacral, or caudal vertebrae are known.

Axis. The axis in *Zarcasaurus* (Figure 29A and B) is identifiable on the basis of the conspicuous dorsolateral orientation of the anterior zygapophyseal facets similar to that in *Araeoscelis*. The axial centrum is elongate, and its midventral keel is sharper than in *Araeoscelis*. The transverse process is positioned low on the side and close to the anterior rim of the centrum, at the anterior end of a long, horizontal ridge. The pedicel of the neural arch is long. Its posterior half has a conspicuous lateral constriction above the level of the centrum. The small portion of the neural arch that remains is indicative of a narrow construction. The extent of the axial neural spine is not known, and the outline presented in the reconstruction (Figure 29A) is based on that in *Araeoscelis*.

Cervical vertebrae. A midcervical and posterior cervical are also present in CM 41704. The former (Figure 29C) is the best preserved of the vertebrae in *Zarcasaurus*. Its structure conforms closely to that in *Araeoscelis*. The centrum is long, with a longitudinal ventral keel that is not as acute as that of the axis. Although no intercentra are known in *Zarcasaurus*, the anterior and posterior faces of the centrum are bevelled for their apparent reception. Like those of the axis, the transverse processes are long oval ridges, bearing a single-headed, posterolaterally directed costal facet. The processes are slightly longer and positioned somewhat higher on the centrum than those of the axis. The neural arch pedicel is long and has a lateral excavation posteriorly.

Although broken, it is clear that the midcervical zygapophyses (Figure 29C) extended well beyond the anterior margin of the centrum, similar to the condition seen in *Araeoscelis* (Vaughn, 1955) and *Petrolacosaurus* (Reisz, 1981). The posterior zygapophyses do not reach beyond the hind end of the centrum. The long, low neural arch is narrow and saddle-shaped in lateral view. The neural spine is also low and elongate. Most of its dorsal edge is chipped away, but it does not appear to be inclined as strongly anteriad as those in *Araeoscelis*.

The neural arch and spine are completely missing in the one preserved posterior cervical vertebra (Figure 29D). The outline reconstruction presented in Figure 29E is conjectural, based on the conditions in *Araeoscelis*. Its serial position may be inferred by the condition of its costal facet. Although holocephalous, a medial constriction subdivides the facet into distinct tubercular and capitular areas, the latter reaching to the anterior rim of the centrum. The centrum is long and sharply keeled. The pedicel is not as long as that in *Araeoscelis*, but closer in proportions to the conditions seen in *Dictybolos* (Olson, 1970: Fig. 5B) and *Petrolacosaurus* (Reisz, 1981: Fig. 14H).

Dorsal vertebrae. The dorsal vertebrae in *Zarcasaurus* (Figure 29F-K) are similar to those in *Araeoscelis*, with some slight proportional differences. The longitudinal and transverse dimensions of the dorsal centra are approximately equivalent, about half those of the cervical length. Dorsal centra have thickened anterior and posterior lips, and their waists are pinched, giving them a rounded midventral ridge. The pedicels of the dorsal neural arches are more clearly defined than those of the cervical vertebrae. The pedicels angle backward, setting the posterior zygapophyses beyond the caudal end of the centrum. The anterior zygapophyses are not well preserved in any of the vertebrae, but they do not appear to have extended significantly beyond the end of the centrum. The zygapophyseal surfaces are worn and apparently were only slightly tilted. Transverse processes are only partially preserved in all of the dorsal vertebrae. It is

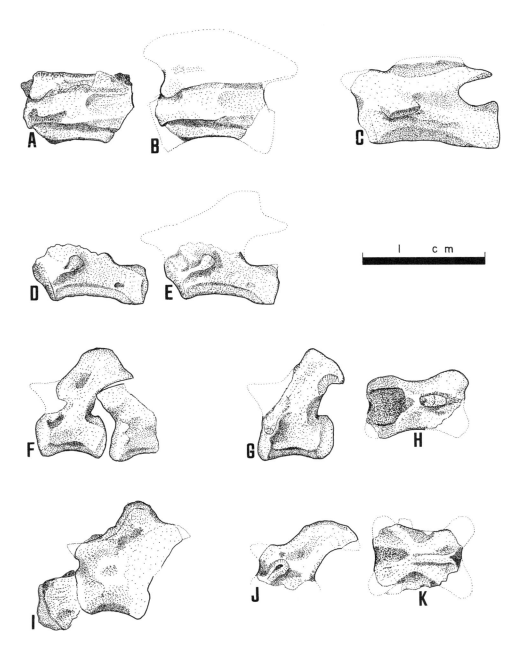

Figure 29. CM 41704, *Zarcasaurus tanyderus*. A-B, portion of axis vertebra and reconstruction. C, midcervical vertebra. D-E, portion of posterior cervical vertebra and reconstruction. F, portions of two dorsal vertebrae in articulation. G-H, left lateral and dorsal aspects of tall-spined dorsal vertebra. I, portions of two dorsal vertebrae in articulation. J-K, left lateral and dorsal aspects of low-spined vertebra. All views are left lateral aspect, except for H and K. Dotted lines indicate estimated extent of original vertebrae, based on proportions observed in the closely related genus *Araeoscelis*.

difficult to determine whether they had separate capitular and tubercular facets, but on one isolated neural arch (Figure 29J) there is a constriction to produce two distinct articular regions.

The neural arches have a slight transverse expansion; as in *Araeoscelis*, they are laterally excavated. They are too worn to allow determination of possible muscular attachments. Evidence for alternation of neural spine height in *Zarcasaurus* is only circumstantial. The presence of alternation in *Araeoscelis* and *Petrolacosaurus*, however, makes the possibility of its existence in *Zarcasaurus* a reasonable supposition. In the more completely preserved dorsals, the neural spines are well-developed and conical in shape. However, one isolated neural arch (Figure 29J and K) possesses a low, narrow, anteroposteriorly elongate spine. While the differences in neural spine structure may be due to serial changes in the vertebrae, the costal facets of the vertebrae illustrated in Figures 29G-H and J-K are not significantly different, and they may be from the same general region of the axial column. Additionally, dorsal views reveal differences in width and anteroposterior length of their neural spines.

PETROLACOSAURUS KANSENSIS

Petrolacosaurus kansensis, from Upper Pennsylvanian sediments near Garnett, Kansas, is the earliest known diapsid reptile. It was first described by Lane (1945) as a sphenacodontid pelycosaur. Stovall et al. (1966) also considered *Petrolacosaurus* to be a pelycosaur, though they allied it with primitive edaphosaurs. Based on materials that should have been assigned to *Petrolacosaurus*, Lane (1946) later described *Podargosaurus hibbardi*, referring it to the Araeoscelidae. Although these initial assignations were incorrect, they underscore the close similarities among *Petrolacosaurus*, araeoscelids, and other groups of primitive amniotes. In 1952, Peabody placed *Podargosaurus hibbardi* in synonymy with *Petrolacosaurus kansensis*.

Abundant and better materials allowed Reisz (1977, 1981) to provide a detailed description of the morphology of *Petrolacosaurus*. He pointed out its diapsid nature, and that the correct assignment of *Petrolacosaurus* was among the Araeoscelidia. More recently, Reisz et al. (1984) have reiterated the close relationship between *Araeoscelis* and *Petrolacosaurus*. As Reisz (1981) has provided a detailed review of the morphology and relationships of *Petrolacosaurus*, its description here will be limited to comments of a comparative nature and a more detailed description of the types of alternation of neural spine structure present. Periodic reference will be made to the illustrations Reisz has provided.

Peabody (1952) estimated that *Petrolacosuarus* had 26 presacral vertebrae, 2 sacrals, and about 60 caudals. Reisz (1981) confirmed these numbers with more complete materials.

Atlas-axis complex. The atlas-axis complex is closely comparable to what elements are known for *Araeoscelis*. The atlantal centrum (Reisz, 1981: Fig. 14D) is fused to the anterodorsal face of the axial intercentrum in mature specimens of *P. kansensis* (KUVP 33607), although the two elements are not always fused in immature individuals (KUVP 1427; Peabody, 1952: Fig. 3). The atlantal intercentrum (Reisz, 1981: Fig. 14B) is squared off and blocky in lateral view and typically crescentic in end view. The atlantal neural arch is a paired structure consisting of a dorsal, posteriorly directed, winglike process and a more ventrally directed process (Reisz, 1981: Fig. 14C and E). Its borders are straight, meeting one another at sharply defined angles. The proatlas (Reisz, 1981: Fig. 14A) is a short, paired structure, similar in general proportions and structure to those of captorhinids and protorothyridids. It is shorter than that in *Varanosaurus* and other pelycosaurs. Peabody (1952) illustrated the proatlas as having been broken, and thus longer than was visible in the available specimens (KUVP 1427, KUVP 33607), but Reisz (1981) restored it with a shorter, more angular anterior terminus. It appears that the reconstruction proposed by Reisz would bring the proatlas into a more appropriate articulation with the exoccipitals of the braincase.

The axis is also closely comparable to that in *Araeoscelis* (Vaughn, 1955: Fig. 5A; Reisz, 1981: Fig. 14E and F), particularly the large, blade-like neural spine and the characteristically dorsolateral orientation of the anterior zygapophyseal facets. However, the pedicel is not as long as that in *Araeoscelis,* being only about two-thirds the length of the centrum.

Cervical vertebrae. Petrolacosaurus has 6 cervical vertebrae, including the specialized atlantal and axial vertebrae, three fewer than in *Araeoscelis*. As in *Araeoscelis*, the cervical vertebrae may be distinguished by their conspicuous elongation. However, the costal articulations remain holocephalous throughout the column; thus, the criteria used by Vaughn (1955) for distinction of cervicals on the basis of the structure of the transverse processes cannot be used in *Petrolacosaurus*.

Cervical vertebrae are elongate in typically araeoscelidian fashion. The cervical neural spines are not as long and low as those in *Araeoscelis* (compare Figure 28A and 28D). Deep lateral excavations are present on the 3rd and 4th neural spines, but are reduced to small depressions on the 5th and 6th. The neural spine of the 3rd cervical is low and triangular in lateral view. More posteriorly, the spines become longer, more squared off and wedge-shaped, and receive a buttressing ridge from the posterior zygapophyses. Mammillary processes are present on the 5th and 6th cervical spines (as well as the anterior dorsals). That of the 5th is directed anteriorly, whereas that of the 6th is on the posterior portion of the neural spine and projects posterolaterally. As in *Araeoscelis*, the mammillary processes are clearly associated with the neural spines and not with the neural arches as in *Eocaptorhinus* (Dilkes and Reisz, 1986).

Dorsal vertebrae. A detailed description of most aspects of the dorsal vertebrae may be found in Reisz (1981). The transverse processes are holocephalous throughout the dorsal series, but by the posteriormost portion of the vertebral column they become short and knobby. The capitular head is lost from the centrum by approximately the 20th presacral, coming to lie on the intercentrum. The pedicels measure about half the central length, and their attachment to the centra remains anteriorly placed through the dorsal series. The dorsal zygapophyses are less tilted than those of the cervicals, only about 10 degrees.

The neural arches are slightly swollen above the posterior zygapophyses. However, their lateral expansion does not reach beyond the transverse processes, and they are not as swollen as in captorhinids, diadectomorph amphibians, or even *Araeoscelis*. The mid- and posterior dorsals exhibit lateral excavations of the neural arch that are deepest among the mid-dorsals (Figure 28E and F).

Most neural spines in *Petrolacosaurus* are tall relative to those of other araeoscelidians. Their anterior borders are inclined posteriorly at an angle of 30-35 degrees to the vertical, but their posterior edges are considerably closer to the vertical axis. Mammillary processes project from the posterior halves of the neural spines of presacrals 7 to 12, and sometimes more posterior neural spines.

Two different patterns of variability in neural spine structure are found in *Petrolacosaurus*. Most specimens assignable to *Petrolacosaurus* (including the most complete example of the axial column, KUVP 9155) display alternation in the anteroposterior length of the neural spine, superimposed on a general decrease in the longitudinal measure of the spine. All of the neural spines in this pattern are tall and fairly slim. Variability in their anteroposterior dimensions is expressed among the anterior dorsals, but in no clearly defined pattern. However, from the 14th presacral on, odd-numbered presacrals are longer in anteroposterior length and more squared off at their dorsal limits. They alternate with spines of a shorter anteroposterior length and a more pointed, tapered tip (Figure 28E and F, respectively).

KUVP 33605 also exhibits alternation in neural spine structure, but in a manner much more like that in captorhinid reptiles. The tall neural spines are typical of those found in most specimens of *P. kansensis*, but they alternate with low, rounded spines, significantly longer in anteroposterior measure. The assignment of KUVP 33605 to the genus *Petrolacosaurus* is supported by the characteristic structure of the lateral fossae of the neural arches. In both tall- and low-spined vertebrae, the fossae are virtually identical to those seen in all other specimens of *Petrolacosaurus*. Reisz (1981) considered KUVP 33605 to be slightly immature, but the neural arches and centra of the vertebrae are firmly fused, and the neural spines are all of smooth, finished bone. There is no indication that this form of alternation may have been a reflection of the juvenile state. As the type of alternation is the only difference evident in the vertebrae of this specimen, the conservative assignment to *Petrolacosaurus* remains appropriate. This second form of alternation is found in presacrals 21 to 26. As with the first example, the most clearly defined pattern of alternation is found just anterior to the pelvic girdle.

Sacral vertebrae and ribs. The sacral vertebrae and ribs are similar to those in *Araeoscelis*, however the lateral fossae that mark the neural arches are more deeply developed. A pair of dorsoventrally directed grooves, which probably accommodated strong intervertebral muscles or ligaments (Reisz, 1981), runs up each side of the anterodorsal edges of the neural spines. The second sacral centrum and intercentrum are not preserved in any of the specimens of *P. kansensis*.

Caudal vertebrae and ribs. Approximately 60 to 65 caudal vertebrae are present in *Petrolacosaurus* (Peabody, 1952; Reisz, 1981). The anterior caudal vertebrae show many of the same characteristics as the posterior dorsals, but are not as robust. The neural spines show no evidence of alternation in structure, either in height or anteroposterior length and they are much smaller than those of the dorsal series. They are inclined posteriorly like those of the anterior dorsals, but are slimmer in transverse section. Diapophyses are present back to the 13th caudal, with ribs on the first seven.

Intercentra. Intercentra are associated with the presacral vertebrae as separate ossifications throughout the presacral column. The only exception is the axial intercentrum which fuses to the atlantal centrum, and then only in mature specimens.

Presacral ribs. Ribs are associated with all of the presacral vertebrae (Reisz, 1981: Fig. 16). Anterior cervical ribs have distinct tubercular and capitular heads, but the 4th, 5th, and 6th, like most of the dorsals following, are single-headed. The cervical ribs are almost straight, directed nearly parallel to the longitudinal axis of the vertebral column. The atlantal ribs taper to a thin point, but the axial rib and those of the following vertebrae become progressively longer and wider. Those more posterior also develop dorsally directed ridges of bone along the proximal halves of their lateral extensions. Like the 7th to 9th cervical ribs in *Araeoscelis,* the 4th and 5th cervical ribs in *Petrolacosaurus* have anterolaterally directed processes. Most of the dorsal ribs (Reisz, 1981: Fig. 16) are very much like those of captorhinids in the shape of their articular surfaces, the curvature and cross-sectional outline of their shafts, and their relative lengths.

ARAEOSCELIDIA: SUMMARY

Detailed patterns of alternation may be described in only two genera of the Araeoscelidia: *Araeoscelis* and *Petrolacosaurus*. In *Araeoscelis* and in one specimen of *Petrolacosaurus*, the structure of the neural spines is like that seen among captorhinids: tall conical spines alternating with low, ridge-shaped, anteroposteriorly elongate spines. Further, the clearest expression of the phenomenon is located primarily among the dorsal vertebrae just anterior to the pelvic girdle, a condition that may be primitive within the Captorhinidae. This might indicate a similar functional scenario, that of an increased potential for dorsiflexion, lateral flexion, and a small amount of lateral rotation. Such a functional scheme is not as easily applied to the pattern more commonly seen in *Petrolacosaurus*. However, it may be conceivable that the alternation of wider and anteroposteriorly elongate neural spines with spines of lesser measure in transverse and longitudinal measures may have allowed space for dorsiflexion that would not have been possible otherwise.

7

DISCUSSION

FUNCTIONAL ANALYSIS

The vertebral column of tetrapods is subject to three basic movements: flexion in the longitudinal-vertical plane (dorsiflexion and ventral flexion), flexion in the horizontal plane (lateral flexion), and rotation (twisting about the longitudinal axis). These movements are usually coupled and rarely occur independently. An analysis of these movements must be considered with reference to two factors: the constraints on movement imposed by the structure of the articular surfaces, and the degree of movement that can be produced by muscular action. Most analyses of the possible vertebral motions in Paleozoic tetrapods have been restricted to the former.

Articular Surfaces of the Vertebrae

The limits and directions of axial motion are dictated by the surface structure and orientation of the articular surfaces of the vertebrae. These include central and zygapophyseal articulations, and accessory articular surfaces such as those found in certain members of the Diadectomorpha.

The centra of most tetrapods act as weight-bearing elements, with the weight of successive vertebral segments transferred to them by the zygapophyses. The centra are constructed to withstand compression, tension, and shear (Wainwright et al., 1976). Their anterior and posterior faces do not appear to have a great deal of influence on the limitations of vertebral movement among the taxa included in this study.

Previous interpretations of axial motion based on zygapophyseal structure and orientation (e.g. Romer, 1956, 1966; Olson, 1976) have considered movements of the vertebral column in only one plane (i.e. only one of the above types of motion). Olson (1936) was the first to systematically address the stabilizing qualities of the vertebral column of Permo-Carboniferous tetrapods, and he later (1976) elaborated on the role of expanded neural arches in lateral stability of the column during locomotion. Only Holmes (1989) has considered the possibility of coupled motions (e.g. lateral bending plus axial rotation) between successive vertebrae in Late Paleozoic tetrapods. The role of alternation in neural spine height may now be included with these interpretations.

Previously, the orientation of zygapophyseal articulations in the groups described in this study was characterized as flat and horizontal (e.g., Olson, 1936, 1976; Romer,

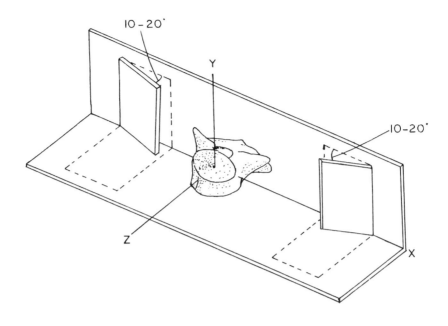

Figure 30. Graphic representation of the zygapophyseal facet inclinations in a typical example of the vertebrae examined in this study. The letters X, Y, and Z represent axes in three-dimensional space. (Diagram adapted from White and Panjabi, 1978.)

1956, 1966), but the orientations associated with vertebrae exhibiting alternation of neural spine height are actually more complex. All show some degree of tilt, sometimes as much as 10 to 20 degrees (Figure 30), and the articular surfaces are often curved. Because of this curvature, the transverse axis of the articular surface of the zygapophysis is taken here to be that line in the transverse plane (Figure 31B) tangent to the center of the zygapophyseal articular surface (Figure 31A). The long axes of the zygapophyses diverge from the sagittal midline.

As the zygapophyses do not lie completely in either horizontal (frontal) or parasagittal planes, movement along their surfaces could not have been strictly limited to either lateral or dorsoventral motions (Figure 31 gives a diagrammatic representation of these planes relative to vertebral structure). Further, their curved surfaces probably allowed a small amount of axial rotation.

The center of rotation of a vertebral segment (Figure 31A) may be defined as the intersection of perpendicular lines drawn from the tangential lines of the transverse axes of the zygapophyses (Gregerson et al., 1967). The articular surfaces may be thought of as arcs on the circumference of a circle, and the perpendiculars as the radii. Because most of the taxa examined in this study possess widely placed zygapophyseal surfaces, the axis of rotation may be located at the level of the neural spine or higher (Figure

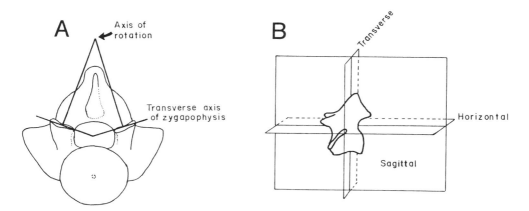

Figure 31. A, representation of means by which the transverse zygapophyseal axes and axis of rotation are determined for a typical captorhinid vertebra. B, diagrammatic representation of transverse, horizontal, and sagittal axes relative to vertebral structure.

31A). The position of the axis of rotation far from the spinal column might seem maladaptive. In fact the movement at the edge of a broad arc described by widely spaced zygapophyses may have approximated movement in a single plane more than rotation about the long axis described by the articiular surfaces of its zygapophyses. However, rotation might have been important to the movement of the vertebra as a whole. Further, the instantaneous axis of rotation is known to shift ventrally with columnar dorsiflexion (Breig, 1960).

Detailed studies of the human vertebral column (summarized in White and Panjabi, 1978) indicate that most vertebral movements are coupled. The only motion that can remain uncoupled is flexion in the plane defined by the curvature of the vertebral column. Wiles (1935) noted that axial rotation must be reduced while at the extremes of dorsal and ventral flexion. If the axial column is thought of as a curved rod, movement in this plane is relatively simple, but motion in planes at an angle to it requires some rotation or distortion of the constituent materials of the rod. Most Late Paleozoic tetrapods possessed only a small amount of spinal curvature, but enough to indicate that lateral flexure was probably coupled with other types of movement.

Though the limits of vertebral motion can be determined by the structure of the zygapophyses, movement can be affected by additional factors. First, the compressive load imposed on the zygapophyses by the weight they bear in quadrupedal animals may reduce movement somewhat. Second, movements may be modified or enhanced by the actions of muscular components.

Muscular Contributions to Movement

As stated previously, few authors have attempted functional interpretations of alternation of neural spine height. Vaughn (1970) originally interpreted it as allowing for the attachment of an "extended nuchal ligament" for support of the large head in *Cap-*

torhinus, but later (1972) retracted this view. Carroll and Gaskill (1978) hypothesized that it allowed ligamentous attachments to every other spine, reducing flexibility and effectively reducing the number of vertebral segments in the long-columned microsaur *Pelodosotis elongatum.* Both proposals involved ligamentous structures, but the evidence presented thus far appears to indicate that alternation of neural spine height was associated with a unique pattern of muscular attachments.

The most conspicuous muscular characteristics in regions of alternation are the configurations of the Mm. interspinales, M. spinalis dorsi, and M. semispinalis dorsi. Although the latter two were surely present in regions of alternation, they do not have well-defined marks of attachment. As they may skip one or more spines between their attachments (Olson, 1936), alternation would not impair their function. However, they may have had less influence in regions of alternation than in regions where alternation is absent, where they appear to have had stronger points of attachment (Figure 10).

Tall-type spines were apparently connected by paired Mm. interspinales. In regions of alternation, the passage of these muscles was facilitated by the intervening low-type spines and, in some cases, furrows lateral to the low-type spine. Such a configuration effectively lengthened the functional interspinous distances, providing a corresponding increase in the range of muscular contraction. Very narrow or ridge-like neural spines (or no neural spine at all, as sometimes found in *Captorhinus* or *Seymouria*), rather than a tall spine, could have allowed for the passage of larger Mm. interspinales. The space otherwise occupied by a taller or wider neural spine could instead provide for a muscle of greater cross-sectional area, with a proportional increase in the force that it would be able to exert.

The movements allowed by the angulation of the zygapophyses, the range of contraction possible due to the longer interspinous distances, and the potential for greater production of force by the proposed muscular system, together indicate that columnar dorsiflexion was a significant component of axial movement among Late Paleozoic tetrapods that exhibited alternation of neural spine height. The presence of small, intervening intercentral elements between successive pleurocentra may have been a means to dissipate tensile stress during such dorsiflexion.

Biomechanical studies of the cervical vertebrae in humans indicate that lateral bending is coupled with axial rotation. This coupling effect causes the neural spines (spinous processes) to rotate outward toward the lateral convexity of the spinal curvature (Penning, 1978; White and Panjabi, 1978). Holmes (1989) has hypothesized that such rotation also occurred in certain Permo-Carboniferous tetrapods.

Whereas lateral flexure of the vertebral column could have produced a coupled axial rotation, the addition of a significant amount of dorsiflexion would have kept the neural spines more closely approximated (closer to the longitudinal-vertical plane). The effective result would have been rotation of the convex side of the vertebral column dorsally, especially each end of the curvature (Figure 32). Further, this would have produced a simultaneous pressure directed toward the contralateral concavity. Relaxation of the muscles of dorsiflexion would have allowed the neural spines to rotate toward the lateral convexity of the column.

Alternation of neural spine height occurs most commonly near the limb girdles. As in *Labidosaurus* and *Captorhinus,* the first sacral vertebra is often very tightly fused to the last presacral. Consequently, alternation may have had a significant effect not only on the movement of the vertebral column, but on the movement of the limb girdles and their associated appendages.

Hildebrand (1976, 1985) has noted that a lateral sequence of footfalls (when diagonal pairs of limbs are providing the power strokes at any particular time; see Figure 33) is the most stable method of locomotion available to lower tetrapods, as the three feet in contact with the substrate outline a more broadly based supporting tripod. An animal utilizing a lateral locomotor sequence has both limbs providing the power stroke on what may be approximated as the concave side of a lateral axial flexure, and the limbs in the recovery phase on the convex side.

Simultaneous dorsiflexion and lateral flexion would prevent rotation of the neural spines toward the lateral convexity of the column. This would help prevent slumping of the backbone while the foot is off the ground, and would also provide an axial contribution to the recovery stroke of the limb. If the greatest amount of lift occurred at each end of the convexity, the lift provided by such dorsiflexion would have been particularly useful at the very beginning of the recovery stroke, when the limb is closest to the posterior end of the curvature.

Simultaneous contralateral pressure toward the concave side of the vertebral column would provide pressure to the limb exerting the power stroke. Holmes (1989) has speculated that the twisting motion produced by outward rotation of the neural spines may have aided locomotion by lowering the foot to the substrate at the beginning of the step cycle. Relaxation of the Mm. interspinales at the end of the recovery stroke would allow such rotation and placement of the foot.

As the hindlimb provides the majority of the forward propulsion, it is not surprising that alternation of neural spine height is most commonly and strongly expressed near the pelvic girdle. Vertebral dorsiflexion of the sacrals and tightly fused ilia was impossible. A concentration and emphasis of alternation just anterior to the pelvic region, and the tight bony coupling of the sacrum with the posteriormost presacrals, may have been sufficient to allow muscular action to influence pelvic motion.

The effects of alternation may have been slightly different anteriorly. Where present, the muscular system associated with alternation probably acted similarly. However, it must be noted that the zygapophyses of anterior dorsal vertebrae are usually more narrowly placed, indicating a lower center of rotation and a greater propensity for axial rotation. Axial rotation was probably difficult to resist. The passive action of foot placement due to outward rotation of the neural spines (Holmes, 1989), plus a small amount of lift due to columnar dorsiflexion, may be a reasonable model for the influence of vertebral motion on the actions of the pectoral limb. Some diadectomorphs are conspicuous in their anterior development of the phenomenon. While the associated locomotory functions hypothesized here is not known to be associated with the pelvic limb in this group, the location of the phenomenon near the pectoral girdle reinforces an interpretation of the importance of alternation to limb mechanics.

Holmes (1977) estimated a 60 degree excursion arc for the pectoral limb in *Captorhinus*, and later (1989) noted that the forelimbs of most Paleozoic tetrapods probably did not possess significantly greater ranges of motion. Similarily, Sumida (1989) estimated the excursion arcs of the pectoral and pelvic limbs in *Labidosaurus* to be ap-

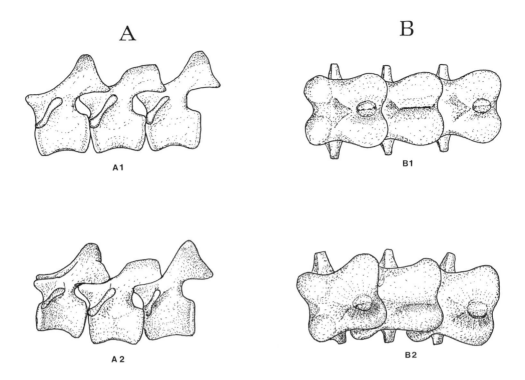

Figure 32. Diagrammatic representation of three vertebrae in articulation, illustrating their relative movements. Column A: left lateral aspect. Column B: dorsal aspect. A1 and B1, vertebrae prior to motion. A2 and B2, relative positions of vertebrae after a combination of lateral flexion and dorsiflexion. Anterior is to the left in both aspects. Vertebral structure based on that of *Captorhinus aguti*.

proximately 50 degrees. Romer (1922, 1956) pointed out that the acetabulum did not restrict excursion of the hindlimb as did the glenoid cavity with the pectoral limb, and Brinkman (1980, 1981) has provided detailed studies of the hindlimb step cycles of extant reptiles. The femur moves through an excursion arc of approximately 75 degrees during the sprawling gait of *Caiman sclerops*, and through an arc of as much as 140 to 165 degrees in *Iguana*. Specimens of *Captorhinus aguti* (UCLA VP 3214) and *Labidosaurus hamatus* (UCLA VP 3600) indicate that the pelvic limb was probably capable of an excursion arc at least as great as that of the pectoral limb, and like extant reptiles, possibly more (Sumida, 1989a). If these estimates are realistic, the aid to the recovery stroke provided by columnar dorsiflexion would have been quite useful. However, more detailed studies of the musculature of the hindlimb, its mechanics, and step cycle in a variety of Paleozoic tetrapods are still necessary. Recently, I have pointed out that captorhinid reptiles probably had well-developed M. caudifemoralis for retraction of the hindlimb (Sumida, 1989a). The well-developed epaxial musculature such as that hypothesized to be associated with alternation of neural spine height also might have aided in stabilizing the pelvic girdle during contraction of the large M. caudifemoralis.

The works of Maurer (1892, 1896, 1899), Furbringer (1900), and Nishi (1916) on the axial structures of lower tetrapods are relevant to this study and may be summarized as follows: in the progression to more advanced forms, the axial musculature is reduced in volume and reorganized; the reorganization is marked by loss of metameric pattern, splitting of the axial musculature into layers, reorientation of the fibers of some of the layers from longitudinal to oblique or transverse, and an increase in the number of segments spanned by some muscles. Olson (1936) refined these studies and added fossil taxa to the analysis. He equated the development of more active terrestrial forms with reduced, more tendinous M. spinalis dorsi and M. semispinalis dorsi in the posterior presacrals of certain Permo-Carboniferous tetrapods. However, he noted that their more anterior portions probably retained their primitive, fleshy configuration. This interpretation is not in conflict with the muscular reconstructions and functional hypotheses presented here for alternation of neural spine height and its prevalent emphasis near the pelvic girdle. Olson (1936) also pointed out that the long lateral muscles of the trunk gradually came to function mainly as flexors of the body in more advanced Paleozoic tetrapods. The present study indicates that 1) alternation of neural spine height and structure probably accommodated a heretofore unknown example of muscles that spanned more than one segment, and 2) the dorsiflexion associated with alternation of neural spine height provided these lateral flexors with a means to make the step cycle more efficient.

Other Mechanical Considerations

Carroll (1986) has speculated that the presence of specialized structures which monitor the passive and active movement in axial musculature is unique to amniotes. Carroll (1986) uses the term stretch receptor for the more restrictive term muscle spindle. Barker et al. (1974) reported that muscle spindles (commonly referred to as stretch receptors in most comparative physiology texts, e.g., Eckert and Randall, 1978) of the somatic musculature are characteristic of tetrapods. Muscle spindles are not found in the axial musculature of extant anamniotes (Barker et al., 1974). However, analogous structures are found in some arthropods, and the appendicular musculature of a wide variety of gnathostomes, including extant chondrichthyans and anurans (Barker et al., 1974). As Barker et al. (1974) refer to analogous structures in vertebrate groups other than amniotes as stretch receptors, the more restricted term muscle spindle is preferred here.

Muscle spindles provide a load-compensating mechanism that helps to refine musculo-skeletal coordination in extant amniotes (Carroll, 1986). Further, alternation of neural spine height appears to be a means to refine and make more efficient the locomotor cycle of taxa included with the Amniota or its close sister groups (phylogenetic scheme of Gauthier et al., 1988). The trend toward increasing adaptations of the axial column and its associated musculature for a higher and more refined degree of terrestriality may all be part of a large, functional complex of anatomical and physiological adaptations. Alternation of neural spine height may have been just a part

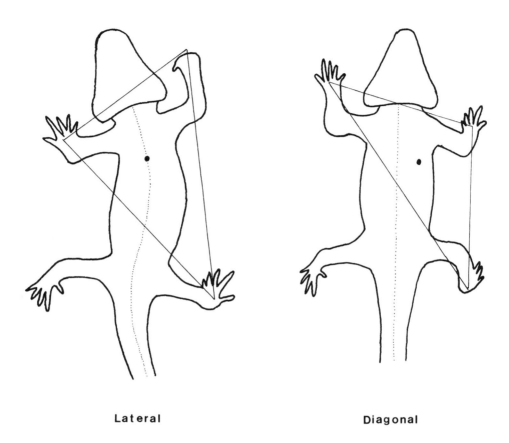

Figure 33. Comparison of factors affecting the stability of a primitive reptile. At left, a larger triangle of support is provided by a lateral sequence of footfalls. The large black spot represents the center of balance. The limbs in the propulsive phase of the stepcycle are located on the concave side of the lateral bend of the vertebral column (dotted line) and those in the recovery phase are on the convex side. At the right, a diagonal sequence does not give as large a triangle of support, and the limbs do not benefit from the interaction of lateral flexion and dorsiflexion. Diagram adapted from Hildebrand (1985).

of the evolution of this whole adaptive complex. Although the microsaurs, seymouriamorphs, and diadectomorphs are amphibians by conventional classification, the development of such structures in these highly terrestrial groups is not inconceivable. Further study of the behavior of muscle spindles of the epaxial musculature of extant reptiles during locomotion would prove useful.

VARIABILITY IN PATTERNS OF ALTERNATION

Although most taxa included in this study seem to fit the functional hypothesis outlined above, certain of them remain enigmatic. As mentioned in Chapter 3, the irregularity of alternation in the Seymouriamorpha may be a partial reflection of a primitive condition wherein alternation is lacking completely. However, its presence even in intermittent regions may have added to columnar stability. The action of larger Mm. interspinales could have provided for extra, actively controlled muscular stabilization of the axial column when one limb was off the ground. This system of additional muscular stabilization of the vertebral column may have been preadapted ("exaptation" of Gould and Vrba, 1982) for use as a locomotory mechanism by virtue of its muscular nature and the proximity of the vertebral column to the limb girdles.

The molgophids and certain long-bodied microsaurs present a distinctly different problem of interpretation. Alternation is quite regular in areas far from the limb girdles in *Pelodosotis, Micraroter*, and *Trihecaton*. It is not at all certain that alternation could have had a significant influence on the action of the limbs in these taxa (as shown previously for *Trihecaton*). However, the ability to flex vertically may have been useful in negotiating a three-dimensionally complex and variable terrain. A similar hypothesis has been advanced to explain the column-long presence of alternation in the captorhinid genus *Labidosaurus* (Sumida, 1987). More functional information on the vertebral column in the Molgophidae is necessary before definitive conclusions can be made about that group.

The study of anatomical structure must not be confined to comparisons of differences in adult structures. Adult structures are end-products of longer ontogenetic processes; i.e., adult structures do not evolve, ontogenies evolve. Central to this is the need to understand the developmental processes that govern the final form taken by morphological structures (Wake, 1970; Waddington, 1975; Maderson, 1975, 1983; Oster and Alberch, 1982; Stephens and Strecker, 1985; Wake and Wake, 1986; Carroll, 1989). Unfortunately, the study of early ontogenetic processes of Paleozoic tetrapods is difficult or impossible in this case. We must look to the developmental processes in extant organisms for clues to possible mechanisms underlying the morphological differences seen in fossils.

Sawin and Hull (1946) observed that vertebral structure in rabbits was not determined segment by segment, but that a small number of specific genetic factors could influence the development of certain characteristics along the entire column. Further, Sawin (1946) showed that the position, structure, and number of vertebral processes could be attributed to just two overlapping developmental factors, each one presumably having a simple basis in differences in the ontogenetic timing that governed vertebral development. Davis (1964) has referred to such developmental factors as "morphogenetic field effects." It may not be unreasonable to postulate that Permo-Carboniferous tetrapods may have had similarly simple and pervasive mechanisms governing the development of their vertebral processes.

The presence of two distinct vertebral "morphs" in the genus *Labidosaurus*, and the variable expression of alternation in *Tseajaia* and *Petrolacosaurus*, remains puzzling. Possible explanations for *Labidosaurus* have included: sexual dimorphism, differing ontogenetic stages, and the possibility that the genus comprises two species characterized by their vertebral structure. The first was interpreted as untestable, the second as untenable, and the third as a difference that could have allowed the sympatric existence of closely related forms by differences in locomotor behavior (Sumida, 1987). This possibility, as well as the distribution of the patterns of alternation among all the groups studied, might also be explained in terms of the potentially homoplastic nature of morphogenetic field effects. Alternatively, the possibility that these differing expressions of vertebral structure could reflect taxonomic distinctions awaits a thorough reanalysis of cranial and appendicular skeletal material in these genera.

PHYLOGENETIC CONSIDERATIONS

The phylogenetic significance of alternation of neural spine structure may be considered at three different levels: whether or not alternation may be a functionally homoplastic feature; the taxonomic levels at which it might be considered a primitive or derived character; and, if it is a phylogenetically useful character, its utility in analysis of the taxa included in this study.

Alternation as a Character

If alternation of neural spine height could be interpreted in terms of morphogenetic fields, its presence throughout the column of certain long-bodied forms (as members of the Ostodolepidae and *Trihecaton*) might be understood as a developmental constraint on all the neural spines required by its presence near the limb girdles. Presumably, such developmental processes were refined in more advanced forms, and alternation was concentrated where its presence was most critical functionally. Such an interpretation argues for alternation as a functionally homoplastic feature.

If alternation is considered to be a character of potential phylogenetic significance, the vertebral structure in the most primitive groups included in this study must be considered. The question of whether the seymouriamorphs or microsaurs are the more primitive grouping is difficult to assess. Microsaurs are known from older sediments, but the phylogenetic relationships of the microsaurs are unclear at best (Vaughn, 1962; Brough and Brough, 1967; Carroll and Baird, 1968; Romer, 1969; Holmes and Carroll, 1977; Carroll and Gaskill, 1978). However, their common primitive ancestor may not be far from the base of reptilian evolution (however, see Panchen and Smithson, 1988 for a dissenting point of view). Traditionally, the Seymouriamorpha have been placed close to the amphibian-reptilian transition (Romer, 1947, 1956, 1966; Heaton, 1980). Carroll (1970) found members of the Gephyrostegoidea to include ancestors of the most primitive reptiles, but more recently, Heaton (1980) and Carroll (1986) have pointed out that the Diadectomorpha might be most appropriately considered as the sister-group of the Reptilia, with the Seymouriamorpha as the sister group of the Diadectomorpha. However, Carroll (1986, 1988) has also indicated that the Diadectomorpha may be too late in time and too derived in some ways to provide a reliable outgroup for the determination of character polarity within the Reptilia. Holmes (1984) has taken issue with a number of the characters upon which Heaton based his analysis, though he was not able to offer an alternative for reptilian origins. Most recently, Gauthier et al. (1988) have also found the Diadectomorpha to be the sister-group of the Amniota, but interpolate *Solenodonsaurus* between the Seymouriamorpha and Diadectomorpha.

Gephyrostegus or *Proterogyrinus* might be considered as reasonable outgroups for analysis of vertebral structure, but their phylogenetic positions have been subject to variable interpretations (for slightly differing viewpoints, see Carroll, 1970; Panchen, 1980; Holmes, 1984). Neither *Gephyrostegus* nor *Proterogyrinus* exhibit swollen neural arches or alternation of neural spine structure. Thus, in a broad sense, a complex of alternation of neural spine height (albeit in an irregular manner in certain seymouriamorphs), together with swollen neural arches, may be a derived characteristic of the Seymouriamorpha, Diadectomorpha, and primitive reptiles.

Phylogenetic Applications

If the combination of swollen neural arches and alternation of neural spine height is taken to be characteristic of the earliest completely terrestrial tetrapods, and if it is considered to have any phylogenetic significance, the fact that it exhibits a variety of permutations requires an explanation of its distribution within the groups described herein. That alternation has not been described for any of the Protorothyrididae, until recently usually accepted as the most primitive reptilian family Carroll, 1969, 1970; Gaffney and McKenna, 1979), is problematic. Alternation may be a homoplastic feature that developed independently in response to similar selective pressures for locomotor efficiency. This interpretation is consistent with: (1) Carroll's (1969, 1970; also Clark and Carroll, 1973) interpretations of the phylogenetically primitive position of the Protorothyrididae within the Reptilia, (2) the questionable phylogenetic position of the Microsauria, and (3) the stratigraphic record, which suggests that the Protorothyrididae may be more primitive than the Captorhinidae.

Alternatively, more traditional hypotheses of primitive reptilian relationships may be subject to re-examination. If the Seymouriamorpha plus Diadectomorpha are accepted as an appropriate outgroup for the Amniota (as has been proposed by Heaton, 1980; Brinkman and Eberth, 1983, 1986), then alternation would be primitive for the Amniota, and the nature of the vertebral column of the Captorhinidae would be interpreted as more primitive than that of the Protorothyrididae. Heaton and Reisz (1986) felt that swollen neural arches, as exemplified by the captorhinids, are primitive for reptiles, and have proposed other characters that may also point to a more derived position for the Protorothyrididae. This interpretation is not consistent with the known stratigraphic occurrences of the protorothyridids, captorhinids, araeoscelids, and *Petrolacosaurus*. However, it must be noted that the temporal distribution of a character is not the primary determinant of polarity in phylogenetic analysis (Hennig, 1966; Stevens, 1980). The more traditionally accepted phylogenies of primitive reptiles (Carroll, 1969, 1970; Gaffney and McKenna, 1979) may be subject to critical re-interpretation as new information becomes available, and that of Heaton and Reisz (1986) might be considered as a reasonable alternative.

Phylogenetic Utility at the Familial Level

Although there is difficulty in using alternation of neural spine height as a character at the supraordinal level and above, its occurrence in most members of certain orders and families might allow it to be useful in the analyses of phylogenetic relations within those groups.

Among many of the constituent members of the Seymouriamorpha and Diadectomorpha, the similarities of neural spine and arch construction, along with the presence of alternation, are consonant with previous interpretations of their close relationship (Romer, 1964; Olson, 1965, 1966; Vaughn, 1964; Moss, 1972; Heaton, 1980). Within that grouping, the more irregular nature of alternation in *Seymouria* appears to be more primitive than its clearer expression near the pectoral girdle among members of the Diadectomorpha.

Alternation appears to be characteristic of the microsaurian family Ostodolepidae as well as the monospecific Trihecatontidae. It may be most useful in identifying members of the Ostodolepidae.

The functional complex of alternation and swollen neural arches is known in only one pelycosaur, *Varanosaurus acutirostris*. It could be interpreted either as a primitively retained character or as an autapomorphic character of *Varanosaurus*. Romer and Price (1940) speculated that the vertebral structure seen in *Varanosaurus* could be a relictual expression of the primitive condition, but Reisz (1972, 1975) has pointed out that the earliest known pelycosaurs have narrow neural arches and blade-like neural spines. Recent analysis of vertebral characters in *Varanosaurus* and other ophiacodont pelycosaurs (Sumida, 1989b) favors the interpretation of alternation in *Varanosaurus* as a character derived in parallel.

Some form of alternation is found in all of the better-known members of the Araeoscelidia. It may be characteristic of that group, and the polarity of its different character-states would seem to have a direct bearing on their interrelationships. Based on a theory of secondary closure of the lower temporal fenestra and its later stratigraphic occurrence, Reisz et al. (1984) interpreted *Araeoscelis* as being more derived than the primitive diapsid *Petrolacosaurus*. The most common expression of alternation, in the height of the neural spines, is found in *Araeoscelis* and some specimens of *Petrolacosaurus*. If the more common form of expression of alternation in neural spine height is accepted as primitive for reptiles, then the differing form of alternation in neural spine width would favor *Petrolacosaurus*, the most gracile of the forms studied here, as more derived than *Araeoscelis*. The increase in stride length suggested by the muscular system associated with alternation may not have been as selectively advantageous in the slender, long-limbed forms. The overall anatomical structure of *Petrolacosaurus*, as the earliest known diapsid (Reisz, 1977, 1980), may be indicative of evolutionary directions taken by reptilian groups more advanced than the protorothyridids and captorhinids. Diapsids may have depended less on axial contributions to locomotion; and the decreasing emphasis on alternation of height of the neural spines in concert with longer, more gracile limbs could be an example of the beginnings of this trend.

Among captorhinids alternation is known in almost all of the better-known forms. Its presence in *Romeria* cannot be ruled out until more complete vertebral remains of this genus are recovered. It is present, but no actual pattern can be established, in *Protocaptorhinus*. The most primitive captorhinid in which a pattern can be confirmed is *Eocaptorhinus*, with alternation most strongly expressed in the posterior region of the presacral column. Other character-states include alternation throughout the column in some examples of *Labidosaurus*, and its presence near only the pectoral and pelvic girdles in most examples of *Captorhinus*. More advanced but fragmentarily represented members of the family appear to show alternation near the limb girdles as well.

SUMMARY

1. Vertebral structure has been described in detail for a variety of Late Paleozoic tetrapods. Many aspects of vertebral structure are quite conservative across a wide variety of groups of primitive tetrapods.

2. A conspicuous exception to the homogeneous vertebral structure previously accepted for Permo-Carboniferous tetrapods is the presence of alternation of neural spine height and neural arch construction.

3. Alternation of neural spine height, and the presence of swollen neural arches and slightly tilted zygapophyses appear to belong to a functional complex common to a wide variety of the earliest truly terrestrial tetrapods. These include: seymouriamorph, diadectomorph, and certain microsaurian amphibians, almost all members of the reptilian family Captorhinidae, the better-known members of the araeoscelidian reptiles, and the pelycosaurian genus *Varanosaurus*.

4. *Seymouria* and *Varanosaurus* display an irregular pattern of alternation. Other groups exhibit more regular patterns of alternation, with occurrence of the phenomenon most often near the limb girdles.

CONCLUSIONS

1. The structure of the zygapophyses indicates that whereas lateral bending occurred in the groups studied, vertebral dorsiflexion was also a major component of movement.

2. The pattern of neural spine alternation appears to have accommodated paired Mm. interspinales that skipped over low-spined vertebrae and attached only to tall-type neural spines. Where present, the phenomenon also included alternating attachments of the M. spinalis dorsi and M. semispinalis dorsi. It is hypothesized that such a muscular configuration would have provided a mechanism for active muscular dorsiflexion of the vertebral column.

3. Concentration of alternation of neural spine height and structure near the limb girdles of many Late Paleozoic tetrapods allowed associated muscular actions to focus their effects on the limbs and their girdles.

4. The ability to flex the vertebral column dorsally would have been useful in preventing rotation of the neural spines toward the convexity of the lateral axial curva-

ture during lateral flexion. The lift thus provided to the convex side of a lateral curvature of the vertebral column near a limb girdle was probably useful in aiding the limb during the recovery stroke of the step cycle. Such flexion also could have provided a small amount of additional pressure to the contralateral limb performing the power stroke.

5. Alternation of neural spine height appears to have been a primitive experiment common to many of the earliest truly terrestrial tetrapods. The phenomenon was apparently a means of making the locomotory step cycle more efficient.

6. The phylogenetic significance of alternation is difficult to assess. It may be a functionally homoplastic feature for which selection pressure was high in a wide variety of taxa because of their common need for efficient terrestrial locomotion. Alternatively, it might be considered a derived characteristic of seymouriamorphs, diadectomorphs, and primitive amniotes. The microsaurs and molgophids do not fit into any clear phylogenetic picture. Acceptance of alternation as a phylogenetic character would require its loss in certain groups, parallel development in certain groups, or a partial re-evaluation of certain of the most commonly accepted hypotheses of the origins and relationships of primitive reptiles.

Appendix

Specimens Examined

Geological Context: Traditional methods of dividing the Lower Permian Formations of north-central Texas have recently been reevaluated by Hentz (1988). Hook (1989) has provided a useful key to the appropriate formational nomenclature for well known collecting localities. This appendix attempts to incorporate the most recent nomenclature wherever possible.

SEYMOURIIDAE

Seymouria agilis Olson

Chickasha Formation, Upper Permian, Blaine County, Oklahoma: UCLA VP 5329; almost complete postcranial skeleton with poorly preserved partial skull and lower jaw.

Seymouria baylorensis Broili

Clear Fork Group undivided, Lower Permian, Baylor County, Texas: FMNH UC663; skull and complete axial column. FMNH UC666; two vertebrae. FMNH UC761; miscellaneous disarticulated and broken elements. FMNH UR368; vertebra. FMNH UR457; separate strings of vertebrae from at least two individuals. FMNH UR458; seven articulated neural arches, two with centra. MCZ 1083; skull, lower jaws, and five dorsal vertebrae. MCZ 1091; two series of three dorsal vertebrae. MCZ 1645; isolated vertebrae. MCZ 1649; disarticulated vertebrae. MCZ 1958; disarticulated skeleton, including vertebrae, skull elements, lower jaws, and appendicular materials. MCZ uncatalogued; string of four dorsal vertebrae and a number of disarticulated individual vertebrae. USNM 9140; almost complete skeleton, portions restored for display. USNM 15553; skull, vertebrae, and portions of limb girdles of two or more individuals. USNM 17047; partial skull, vertebrae, ribs and portion of pectoral girdle. USNM 21902; almost complete skeleton.

Clear Fork Group undivided, Lower Permian, Taylor County, Texas: UCLA VP 456; dorsal vertebra.

Seymouria grandis Olson

Hennessey Formation, Lower Permian, Fairmont County, Oklahoma: UCLA VP 3152; partial skull, vertebrae.

Clear Fork Group undivided, Lower Permian, Taylor County, Texas: UCLA VP 458; dorsal vertebrae. UCLA VP 559; dorsal vertebrae. UCLA VP 570; anterior dorsal vertebrae. UCLA VP 571; vertebrae. UCLA VP 572; dorsal vertebra. UCLA VP 613; partial skeleton. UCLA VP 3521; vertebrae.

Seymouria sanjuanensis Vaughn

Cutler Formation, Lower Permian, Rio Arriba County, New Mexico: CM 28596; nearly complete articulated skeleton, missing forelimbs and most of hindlimbs. CM 28597; nearly complete articulated skeleton, missing forelimbs and most of hindlimbs. CM 28598; most of skull, pectoral girdle and presacral vertebral column in articulation. CM 28599; complete vertebral column and portion of pelvic girdle. CM 34900; pelvic girdle, posterior dorsal vertebrae, sacral vertebrae, and anterior caudal vertebrae. CM 38022; nearly complete skeleton with tail, missing portions of forelimbs and hindlimbs.

Organ Rock Shale, Cutler Group, San Juan County, Utah: NTM 1025; impression and bone of skull roof, anterior vertebrae, shoulder girdle, and parts of front limbs. NTM 1026; impression and small amount of bone of skull roof, anterior vertebrae, parts of front limbs. NTM 1029; posterior, sacral, and caudal vertebrae, femur, tibia, and fibula. NTM 1030; string of caudal vertebrae. NTM 1031; partial shoulder girdle and attached proximal portion of humerus. NTM 1033; partial braincase, fragment of palate, portions of dermal skull roof, complete vertebral column excepting some caudals, partial pectoral and pelvic girdles with limbs. NTM 1034; portion of dermal skull roof and right mandible, fragment of palate, most of presacral vertebral column, partial pectoral and pelvic girdles, parts of pelvic limbs. NTM 1035; portions of palate and lower jaws, part of right pectoral girdle and limb, anterior portion of vertebral column. NTM 1036; vertebral column with parts of limb girdles and free limbs. NTM 1037; vertebral column with parts of limb girdles and free limbs. NTM 1037; vertebral column with parts of limb girdles and free limbs.

LIMNOSCELIDAE

Limnoscelis paludis Williston

Cutler Formation, Lower Permian, El Cobre Canyon, Rio Arriba County, New Mexico: FMNH UR306; two articulated strings of seven vertebrae and other scattered materials. YPM 811; complete skeleton.

Limnoscelis sp.

Sangre de Cristo Formation, Upper Pennsylvanian, Fremont County, Colorado: CM 47653; partially disarticulated and scattered skeleton of a single individual.

TSEAJAIIDAE

Tseajaia campi Vaughn
Organ Rock Shale, Cutler Group, Lower Permian, San Juan County, Utah: UCMP V24216/63841; string of seven articulated dorsal vertebrae. UCMP V4225/59102; complete skeleton.

Tseajaia sp.
Cutler Formation, Lower Permian, Arroyo de Agua, Rio Arriba County, New Mexico: CM 38033; skull, vertebral column, part of the pectoral girdle and associated limbs, portion of the hindlimbs. CM 38042; skull, almost complete vertebral column, some limb material.

DIADECTIDAE

Diadectes sideropelicus Cope
Petrolia Formation, Lower Permian, Archer County, Texas: FMNH UR27; skull, vertebrae, pectoral girdle.
Petrolia Formation, Lower Permian, Baylor County, Texas: FMNH UC1177; skull, vertebrae, pectoral girdle.

Diadectes tenuitectus (Cope) Clear Fork Group undivided, Lower Permian, Coffee Creek, Baylor County, Texas: FMNH UC 654; posterior half of presacral column, first sacral vertebra.

Diadectes sp.
Clear Fork Group undivided, Lower Permian, Middle Coffee Creek, Baylor County, Texas: FMNH UC1075; complete presacral column, pectoral and pelvic girdles, sacrum, anterior portion of tail. (Proximity to Coffee Creek questionable): UCLA VP 3732; vertebra.
Waggoner Ranch Formation(?), Lower Permian, near Seymour, Baylor County, Texas: FMNH UC 1234; vertebrae, scrap. FMNH UC1235; vertebrae, scrap.
Cutler Formation, Lower Permian, San Juan County, Utah: UCLA VP 2962; vertebra.
Archer City(?) Formation, Lower Permian, Archer County, Texas: UCLA VP 344; two articulated strings of four and six vertebrae.
Organ Rock Shale, Cutler Group, San Juan County, Utah: NTM 1002; two long strings of dorsal vertebrae with ribs, parts of forelimbs, strings of caudal vertebrae, left lower jaw, parts of skull, teeth. NTM 1004; vertebra.

Diadectes(?) sp.
Cutler Formation, Lower Permian, Arroyo de Agua, Rio Arriba County, New Mexico: CM 38036; juvenile, incomplete skull and almost complete postcranial skeleton.

MICROSAURIA

Pantylus cordatus Cope

Waggoner Ranch Formation, Lower Permian, Baylor County, Texas: FMNH UC1069; two partial skulls, vertebrae and ribs, pectoral and pelvic girdles, limb elements.

Archer City Formation, Lower Permian, Archer County, Texas: MCZ 3302; skull and anterior portion of skeleton. UT 40001-1; vertebral column, limb girdles, hind limb, and questionably associated skull (sectioned by Romer, 1969). UT 40001-6; skull, vertebral column, forelimb. UT 40001-7; skull, vertebral column, other scattered associated material. UT 40001-8; disarticulated material from a number of individuals.

Ostodolepis brevispinatus Williston

Clear Fork Group undivided, Lower Permian, West Coffee Creek, Wilbarger County, Texas: FMNH UR680; seven dorsal vertebrae, associated ribs and scales.

Micraroter erythrogeios Daly

Hennessey Formation, Lower Permian, Tillman County, Oklahoma: FMNH UR2311; skull with jaw, four neural arches and two centra, portion of pectoral girdle.

Trihecaton howardinus Vaughn

Sangre de Cristo Formation, Upper Pennsylvanian, Fremont County, Utah: UCLA VP 1743; maxilla, mandible, complete presacral column, portions of limb girdles and associated limb elements, scales. UCLA VP 1744; twelve caudal vertebrae and haemal arches, scales as in 1743.

MOLGOPHIDAE

Molgophis macrurus (Cope)

Linton Diamond Coal Mine, Upper Freeport coal of the Allegheny Group, Middle Pennsylvanian, Jefferson County, Ohio: MCZ 3327 (latex peel of AMNH 6840); string of ten dorsal vertebrae and associated ribs.

Pleuroptyx clavatus Cope

Linton Diamond Coal Mine, Upper Freeport coal of the Allegheny Group, Middle Pennsylvanian, Jefferson County, Ohio: MCZ 3327 (latex peel of AMNH 6863); string of twelve dorsal vertebrae and associated ribs.

PELYCOSAURIA

Varanosaurus acutirostris Broili

Clear Fork Group undivided, Lower Permian, Baylor County, Texas: AMNH, 4174; partial skull and majority of postcranial skeleton. BSPHM 1901 XV 20; nearly complete skeleton.

Clear Fork Group quivalent, Lower Permian, Garvin County, Oklahoma: FMNH PR1670; complete skull and associated atlas-axis complex.

CAPTORHINIDAE

Not included in this list are a large number of isolated vertebrae or short strings of vertebrae, most uncatalogued, assignable to *Captorhinus, Eocaptorhinus*, or *Labidosaurus*. Uncatalogued materials examined include those in the collections of the

Appendix: Specimens Examined

Field Museum of Natural History, Chicago; the University of Kansas Museum of Natural History, Lawrence; the University of California Museum of Paleontology, Berkeley; and the University of California, Los Angeles Vertebrate Paleontology collection.

Romeria primus Clark and Carroll

Archer City Formation, Lower Permian, Cottonwood Creek, Archer County, Texas: MCZ 1963; skull, anterior presacral vertebrae, partial forelimb.

Romeria texana Clark and Carroll

Archer City Formation, Lower Permian, Archer City Bone Bed, Archer County, Texas: MCZ 1480; dermal skull roof, cheek and palate.

Protocaptorhinus pricei Clark and Carroll

Nocona Formation, Lower Permian, Rattlesnake Canyon, Archer County, Texas: MCZ 1478; skull and portion of anterior postcranial skeleton.

Wellington Formation, Lower Permian, Orlando, Logan County, Oklahoma: UCLA VP 3531; vertebrae, femur, scraps. UCLA VP 3537; vertebrae and ribs. UCLA VP 3539; vertebrae, partial femur, scrap. UCLA VP 3541; vertebrae, partial maxilla, and limb elements. UCLA VP 3545; vertebrae and postcranial fragments. UCLA VP 3546; vertebrae and limb fragments. UCLA VP 3622; vertebrae. UCLA VP 3626; vertebrae, teeth, cranial fragments. UCLA VP 3630; vertebrae and partial limb bone.

Eocaptorhinus laticeps (Williston)

Waggoner Ranch Formation, Lower Permian, Mitchell Creek, Baylor County, Texas: FMNH UC642; complete skull and majority of postcranial skeleton.

Cutler Formation, Lower Permian, Rio Arriba County, New Mexico: FMNH UR735; various fragments, including eight dorsal vertebrae exposed in lateral and ventral view.

Wellington Formation, Lower Permian, McCann Rock Quarry, 2 miles northeast of Eddy, Kay County, Oklahoma: OUSM 15020A; anterior protion of skull. OUSM 15020B; posterior portion of skull, anterior 21 vertebrae, and partial forelimb. OUSM 15024; partial skull roof, partial vertebral column, manus, and pes. OUSM 15025; fragments of skull, limbs, and vertebrae. OUSM 15101; complete skull and worn anterior half of vertebral column.

Labidosaurus hamatus Cope

Clear Fork Group undivided, Lower Permian, Baylor County, Texas: AMNH 4417; mandible, partial postcranial skeleton including vertebrae 11-25, pelvic girdle, and sacral vertebrae. FMNH P12758; vertebrae and pectoral girdle. FMNH UC174; skull and articulated postcranial skeleton. FMNH UC178; badly crushed skull and most of axial skeleton. FMNH UC726; vertebrae, humeri, and femur. FMNH UC730; vertebrae, fragments of jaws, and other bones. FMNH UR161; vertebrae and limb fragments. MCZ 8923; string of nine articulated mid-dorsal vertebrae. UCLA VP 436; disarticulated vertebrae, limb elements, and fragments. UCLA VP 3167; skull and nearly complete postcranial skeleton. UCLA VP 3200; skull and nearly complete postcranial skeleton. UCLA VP 3491; juvenile, articulated tail and fragments. USNM 17045; skull and complete postcranial skeleton.

Captorhinus aguti (Cope)

Clear Fork Group undivided, Lower Permian, Baylor County, Texas: AMNH 4332; almost complete skeleton. MCZ 1059; skull, anterior dorsal vertebrae, and scattered fragments. MCZ uncatalogued; presacral vertebrae 22-25, sacral, caudal and other scattered vertebrae, and pelvic girdle.

Clear Fork Group equivalent, Dolese Brothers Quarry (Fort Sill locality), Lower Permian, Comanche County, Oklahoma: AMNH 4434; jaws, skull, dorsal vertebrae, and partial pectoral girdle. AMNH 5954; skull and almost complete vertebral column. AMNH uncatalogued; skull, vertebral column, partial pectoral and pelvic girdles, limb elements, and skin impression. KUVP 9609; atlantal neural arch. KUVP 9610; atlantal centrum. KUVP 9978; complete skull. OUSM 15133; 3 articulated strings of vertebrae. OUSM 15136; jaw and short articulated string of vertebrae. OUSM 15146; partial vertebral column and hind limb. UCLA VP 1735; articulated string of three dorsal vertebrae. UCLA VP 3752; atlantal centrum. UCLA VP 3753; atlantal centrum. UCLA VP 3760; tip of tall-type neural spine, frontal section. UCLA VP 3761; tall-type neural spine, frontal section. UCLA VP 3762; tall-type neural spine, frontal section. UCLA VP 3763; tall-type neural spine, frontal section. UCLA VP 3764; low-type neural spine, frontal section. UCLA VP 3765; low-type neural spine, frontal section. UCLA VP 3766; tall-type neural spine, frontal section. UCLA VP 3767; low-type neural spine, frontal section. UCLA VP 3768; tall-type neural spine, frontal section. UCLA VP 3769; low-type neural spine, sagittal section. UCLA VP 3770; tall-type neural spine, sagittal section. UCLA VP 3771; tip of tall-type neural spine, frontal section. UCLA VP 3772; tall-type neural spine, frontal section. UCLA VP 3773; posterior portion of tall-type neural spine, frontal section.

Arroyo Formation, Lower Permian, Granfield, Oklahoma: UCLA VP 754; skull, jaws, and postcranial parts. UCLA VP 3176; partial skeleton and nodule with bone. UCLA VP 3214; almost complete skeleton.

?*Captorhinus aguti* (Cope)

Clear Fork Group equivalent, Dolese Brothers Quarry (Fort Sill locality), Lower Permian, Comanche County, Oklahoma: USNM 391917; partial skeleton, including humerus, pectoral girdle, and anterior portion of the vertebral column.

Captorhinikos chozaensis Olson

Hennessey Formation, Lower Permian, Cleveland County, Oklahoma: FMNH UR857; partial skull and skeleton. FMNH UR858; partial vertebral column, ribs, and limb elements. FMNH UR859; partial skull and skeleton. UCLA VP 3058; vertebra. USNM 21275; skull and partial skeleton.

Captorhinikos parvus Olson

Hennessey Formation, Lower Permian, Cleveland County, Oklahoma: UCLA VP 2895; partial skeleton including partial articulated vertebral column. UCLA VP 2896; partial skeleton including partial articulated vertebral column. UCLA VP 2904; vertebrae and associated scraps.

Captorhinikos valensis Olson

Clear Fork Group undivided, Lower Permian, Knox County Texas: FMNH UR106; six presacral vertebrae with ribs, impression of about five more anterior to these, and scattered cranial materials. FMNH UR107; seven presacral vertebrae.

Kahneria seltina Olson

San Angelo Formation, Upper Permian, Driver Ranch, Knox County, Texas: FMNH UR562; partial jaw, partial postcranium and scrap.

Labidosaurikos sp.

Clear Fork Group undivided, Lower Permian, Baylor County Texas: FMNH UR15; dorsal vertebra. FMNH UR225; dorsal vertebra. FMNH UR373; string of three dorsal vertebrae.

Clear Fork Group undivided, Lower Permian, Taylor County, Texas: UCLA VP 577; string of six dorsal vertebrae with portions of ribs.

Rothianiscus multidonta (Olson and Beerbower)

San Angelo Formation, Upper Permian, Pease River, Hardeman County, Texas: FMNH 129; partial skull and anterior portion of vertebral column with associated postcranial materials. FMNH UR131; four dorsal vertebrae and associated scrap. FMNH UR263; posterior dorsal vertebrae and associated limb elements.

Chickasha Formation, Upper Permian, Blaine County, Oklahoma: UCLA VP 3673; vertebra and part skull.

ARAEOSCELIDAE

Araeoscelis sp.

Clear Fork Group undivided (Admiral Formation), Lower Permian, Baylor County, Texas: USNM 22078; axis and cervical vertebrae.

Clear Fork Group undivided (Belle Plains Formation), Lower Permian, Baylor County, Texas: FMNH UC659; posterior portion of postcranial skeleton. FMNH UC660; pectoral girdle, posterior dorsal vertebrae, pelvic girdle, anterior and posterior limbs. FMNH UC661; portions of two skulls, associated with UC659, 660, 662. FMNH UC662; most of postcranial skeleton, including tarsi. MCZ 2304; portions of two skulls, including jaw fragments, vertebrae and ribs, portions of each limb girdle, and associated limb elements. MCZ 4383; partial skull and jaw, vertebrae and limb elements.

Zarcasaurus tanyderus Brinkman, Berman and Eberth

Cutler Formation, Lower Permian, Arroyo de Agua, Rio Arriba County, New Mexico: CM 41704; portion of lower jaw, disarticulated vertebrae, disarticulated limb elements.

ARAEOSCELIDAE (?)

Dictybolos tener Olson

Wellington Formation, Lower Permian, Logan County, Oklahoma: FMNH 1041; partial skull, jaws, partial vertebral column, girdle and limb elements. FMNH 1045; dorsal vertebra. FMNH 1046; dorsal vertebra. FMNH 1052; dorsal vertebra. FMNH 1054; cervical vertebra. FMNH 1059; sacral vertebra. FMNH 1060; anterior caudal vertebrae. FMNH 1067; maxilla, vertebra, rib, other fragments. FMNH 1073; partial skull and caudal vertebrae. FMNH 1075; cervical vertebra and rib. FMNH 1079, caudal vertebrae. FMNH 1101; parts of approximately 25-30 vertebrae. FMNH 1108; caudal vertebra. FMNH 1140 skull scrap and vertebra.

PETROLACOSAURIDAE

Petrolacosaurus kansensis Lane

Garnett locality (correlated with Rock Lake Shale, Platte Valley, southeastern Nebraska; type locality), Upper Pennsylvanian, Anderson County, Kansas: KUVP 9951; partial skull and most of postcranial skeleton. KUVP 9956; nearly complete vertebral and costal series except for atlas, axis, and distal caudals. KUVP 12483; possible juvenile specimen of *Petrolacosaurus(?)*. KUVP 33605; posterior portion of vertebral column. KUVP 33606; partial skull and almost complete postcranial skeleton. KUVP 33607; partial skull and anterior portion of postcranium.

Literature Cited

Baird, D.
1965. Paleozoic lepospondyl amphibians. American Zoologist, 5:287-294.

Baird, D., and R. L. Carroll
1967. *Romeriscus*, the oldest known reptile. Science, 157:56-157:56-59.

Barker, D., C. C. Hunt, and A. K. McIntyre
1974. Muscle receptors. *In* C. C. Hunt, ed., *Handbook of Sensory Physiology*, Vol. 3(2), pp. 1-299. Berlin and New York: Springer-Verlag.

Berman, D. S, D. A. Eberth, and D. B. Brinkman
1988. *Stegotretus agyrus* a new genus and species of microsaur (amphibian) from the Permo-Pennsylvanian of New Mexico. Annals of Carnegie Museum, 57:293-323.

Berman, D. S, and R. R. Reisz
1986. Captorhinid reptiles from the Early Permian of New Mexico, with description of a new genus and species. Annals of Carnegie Museum, 55:1-28.

Berman, D. S, R. R. Reisz, and D. A. Eberth
1987a. A new genus and species of trematopid amphibian from the Late Pennsylvanian of north-central New Mexico. Journal of Vertebrate Paleontology, 7:252-269.

1987b. *Seymouria sanjaunensis* (Amphibia, Batrachosauria) from the Lower Permian Cutler Formation of north-central New Mexico and the occurrence of sexual dimorphism in that genus questioned. Canadian Journal of Earth Sciences, 24:1769-1784.

Bolt, J. R., and R. DeMar
1975.   An explanatory model of the evolution of multiple rows of teeth in *Captorhinus aguti*. Journal of Paleontology, 49:814-832.

Breig, A.
1960.   *Biomechanics of the Central Nervous System: Some Basic Normal and Pathological Phenomena*. Stockholm: Almquist and Wiksell; 183 pages.

Brinkman, D.
1980.   The hind limb step cycle of *Caiman sclerops* and the mechanics of the crocodile tarsus and metatarsus. Canadian Journal of Zoology, 58:2187-2200.
1981.   The hind limb step cycle of *Iguana* and primitive reptiles. Journal of Zoology, 181:91-103.

Brinkman, D., D. S Berman, and D. A. Eberth
1984.   A new araeoscelid reptile, *Zarcasaurus tanyderus*, from the Cutler Formation (Lower Permian) of north-central New Mexico. New Mexico Geology, 6:34-39.

Brinkman, D., and D. A. Eberth
1983.   The interrelationships of pelycosaurs. Breviora, no. 473; 35 pages.
1986.   The anatomy and relationships of *Stereophallodon* and *Baldwinonus* (Reptilia, Pelycosauria). Breviora, no. 485; 34 pages.

Broili, F. von
1904.   Permische Stegocephalen und Reptilien aus Texas. Paleontographica, 51:1-120.
1914.   Über den Schädelbau von *Varanosaurus acutirostris*. Separat- Abdruck aus dem Centralblatt f. Min. no. 1:26-29.

Broom, R.
1913.   On the structure and affinities of *Bolosaurus*. Bulletin of the American Museum of Natural History, 32:509-516.

Brough, M. C., and J. Brough
1967.   Studies on early tetrapods I. The Lower Carboniferous microsaurs, II. *Microbrachis*, the type microsaur, III. The genus *Gephyrostegus*. Philosophical Transactions of the Royal Society of London, 252:107-165.

Bystrow, A. P.
1944.   *Kotlassia prima* Amalitzky. Bulletin of the Geological Society of America, 55:1-28.

Literature Cited

Carroll, R. L.
1967. A limnoscelid reptile from the Middle Pennsylvanian. Journal of Paleontology, 41:1256-1261.
1968. The postcranial skeleton of the Permian microsaur *Pantylus*. Canadian Journal of Zoology, 46:1175-1192.
1969. Origin of reptiles. *In* C. Gans, A. d'A. Bellairs and T. S. Parsons, eds., *Biology of the Reptilia*, vol. 1, Morphology A, pp. 1-44. New York: Academic Press.
1970. The ancestry of reptiles. Philosophical Transactions of the Royal Society of London, ser. B, 257:267-308.
1986. The skeletal anatomy and some aspects of the physiology of primitive reptiles. *In* N. Hotton III, P. D. MacLean, J. J. Roth, and E. C. Roth, eds., *The Ecology and Biology of Mammal-like Reptiles*, pp. 25-46. Washington, D.C.: Smithsonian Institution Press.
1988. *Vertebrate Paleontology and Evolution*. New York: W. H. Freeman and Sons. 698 pages.
1989. Developmental aspects of lepospondyl vertebrae in Paleozoic tetrapods. Historical Biology, 3:1-25.

Carroll, R. L., and D. Baird
1968. The Carboniferous amphibian *Tudianus* [*Eosauravus*] and the distinction between reptiles and microsaurs. American Museum Novitates, no. 2337; 50 pages.
1972. Carboniferous stem-reptiles of the family Romeriidae. Bulletin of the Museum of Comparative Zoology, 143:321-364.

Carroll, R. L., and P. Gaskill
1978. The Order Microsauria. Memoirs of the American Philosophical Society, 126; 211 pages.

Case, E. C.
1902. On some vertebrate fossils from the Permian beds of Oklahoma. *In* A. H. Fleet, *Second Biennial Report, Department of Geology and Natural History, Territory of Oklahoma*, pp. 62-68.
1910. New or little known reptiles and amphibians from the Permian (?) of Texas. Bulletin of the American Museum of Natural History, 28:163-181.
1911. A revision of the Cotylosauria of North America. Publications of the Carnegie Institute of Washington, no. 415; 122 pages.
1917. The skeleton of *Poecilospondylus francisi*, a new genus and species of Pelycosauria. Bulletin of the American Museum of Natural History, 28:183-188.
1929. Description of a nearly complete skeleton of *Ostodolepis brevispinatus* Williston. Contributions from the Museum of Paleontology, University of Michigan, 3:81-107.

Case, E. C., and S. W. Williston
1913. Description of a nearly complete skeleton of *Diasparactus zenos*. In E. C. Case, S. W. Williston, and M. G. Mehl, eds., Permo-Carboniferous vertebrates from New Mexico, pp. 17-35. Carnegie Institute of Washington Publication 181.

Clark, J., and R. L. Carroll
1973. Romeriid reptiles from the Lower Permian. Bulletin of the Museum of Comparative Zoology, 144:353-407.

Cope, E. D.
1868. Synopsis of the extinct Batrachia of North America. Proceedings of the Academy of Natural Sciences of Philadelphia, 1868:208-221.
1875. Synopsis of the extinct Batrachia from the Coal Measures. Report of the Geological Survey of Ohio, 2:349-411.
1881. On some new Batrachia and Reptilia from the Permian beds of Texas. United States Geological Survey Territory Bulletin, 6:79-82.
1882   Third contribution to the history of the Vertebrata from the Permian formation of Texas. American Philosophical Society Proceedings, Philadelphia, 20:447-461.
1886. Systematic catalogue of the species of vertebrates found in the beds of the Permian epoch in North America with notes and descriptions. American Philosophical Society Transactions, 16:285-297.
1892. On the homologies of the posterior cranial arches in the Reptilia. Transactions of the American Philosophical Society, 17:11-26.
1896. *Labidosaurus* Cope. Proceedings of the American Philosophical Society, p. 185.

Credner, H.
1889. Die Stegocephalen und Saurier aus dem Rothliegenden des Plauen'schen Groundes bei Dresden: VIII. Theil. *Kadaliosaurus priscis* Cred. Deutsche geologische Gesellschraft, Zeitschrift, 41:319-342.

Daly, E.
1973. A Lower Permian vertebrate fauna from southern Oklahoma. Journal of Paleontology, 46:562-589.

Davis, D. D.
1964. The giant panda, a morphological study of evolutionary mechanisms. Fieldiana: Zoology Memoirs, 3:1-339.

Dilkes, D. W., and R. R. Reisz
1986. The axial skeleton of the Early Permian reptile *Eocaptorhinus laticeps* (Williston). Canadian Journal of Earth Sciences, 23:1288-1296.

Eckert, R., and D. Randall
1978. *Animal Physiology*. San Francisco: W. H. Freeman and Company. 557 pages.

Efremov, I. A.
1946. On the subclass Batrachosauria, a group of forms intermediate between amphibians and reptiles. Isv. Biol. Div. Sci. U.S.S.R., 6:615-638. (In Russian; English summary.)

Enlow, D. H.
1969. The bone of reptiles. *In* C. Gans, A. d'A. Bellairs and T. S. Parsons, eds., *Biology of the Reptilia*, Vol. 1, Morphology A, pp. 45-80. New York: Academic Press.

Enlow, D. H., and S. O. Brown
1957. A comparative histological study of fossil and recent bone tissues, Part II. Texas Journal of Science, 9:186-214.

Evans, F. G.
1939. The morphology and functional evolution of the atlas-axis complex from fish to mammals. Annals of the New York Academy of Sciences, 39:29-104.

Fox, R. C., and M. C. Bowman
1966. Osteology and relationships of *Captorhinus aguti* (Cope) (Reptilia: Captorhinomorpha). University of Kansas Paleontological Contributions, Vertebrata, article 11; 79 pages.

Furbringer, M.
1900. Zur ver gleichenden Anatomie der Brustchulterapparates und der Schultermuskeln. IV Theil. Jenaische Zeitschrift, 34:215-718.

Gaffney, E. S., and M. C. McKenna
1979. A late Permian captorhinid from Rhodesia. American Museum Novitates, no. 2688; 15 pages.

Gauthier, J. A., A. G. Kluge, and T. Rowe
1988. The early evolution of the Amniota. *In* M. J. Benton, ed., Systematics Association Special Volume no. 35A *The Phylogeny and Classification of the Tetrapods, Volume 1: Amphibians, Reptiles, Birds*, pp. 103-155. Oxford: Clarendon Press.

Gould, S. J., and E. S. Vrba
1982. Exaptation-a missing term in the science of form. Paleobiology, 8:4-15.

Gregerson, G. G., and D. B. Lucas
1967. An *in vivo* study of the axial rotation of the human thoracolumbar spine. Journal of Bone and Joint Surgery, 49-A:247-262.

Gregory, J. T., F. E. Peabody, and L. I. Price
1952. Revision of the Gymnarthridae American Permian microsaurs. Peabody Museum of Natural History Bulletin, 10:1-77.

Heaton, M. J.
1979. Cranial anatomy of primitive captorhinid reptiles from the late Pennsylvanian and early Permian, Oklahoma and Texas. Oklahoma Geological Survey Bulletin 127; 84 pages.
1980. The Cotylosauria: a reconsideration of a group of archaic tetrapods. *In* A. L. Panchen, ed., Systematics Association Special Volume no. 15: *The Terrestrial Environment and the Origin of Land Vertebrates*, pp. 497-551. New York: Academic Press.

Heaton, M. J., and R. R. Reisz
1980. A skeletal reconstruction of the early Permian captorhinid reptile *Eocaptorhinus laticeps* (Williston). Journal of Paleontology, 54:136-143.
1986. Phylogenetic relationships of captorhinomorph reptiles. Canadian Journal of Earth Sciences, 23:402-418.

Hennig, W.
1966. *Phylogenetic Systematics* (trans. D. D. Davis and R. Zangerl). Urbana, Illinois: University of Illinois Press. 263 pages.

Hentz, T. F.
1988. Lithostratigraphy and paleoenvironments of Upper Paleozoic continental red beds, north-central Texas: Bowie (new) and Wichita (Revised) Groups. Texas Bureau of Eonomic Geology Report of Investigations No. 170. 55 pages + plate insert.

Hildebrand, M.
1976. Analysis of tetrapod gaits: general considerations and symmetrical gaits. *In* R. M. Herman, ed., *Neural Control of Locomotion, Advances in Behavioral Biology*, 18. New York: Plenum Press.
1985. Walking and running. *In* M. Hildebrand, D. M. Bramble, K. F. Liem and D. B. Wake, eds., *Functional Vertebrate Morphology*, pp. 38- 57. Cambridge, Massachusetts.: The Belknap Press of Harvard University Press.

Hilton, J.
1942. Thumbnail in the book of time. Desert Magazine, 5:18-22.

Holmes, R.
1977. The osteology and musculature of the pectoral limb of small captorhinids. Journal of Morphology, 152:101-140.
1980. *Proterogyrinus scheelei* and the early evolution of the labyrinthodont pectoral limb. *In* A. L. Panchen, ed., Systematics Association Special Volume no. 15: *The Terrestrial Environment and the Origin of Land Vertebrates.* pp. 351-376. New York: Academic Press.
1984. The Carboniferous amphibian *Proterogyrinus scheeli* Romer, and the early evolution of tetrapods. Philosophical Transactions of the Royal Society of London, 306:431-524.
1989. Functional interpretations of the vertebral structure in Paleozoic amphibians. Historical Biology, 2:111-124.

Holmes, R., and R. Carroll
1977. A temnospondyl amphibian from the Mississippian of Scotland. Bulletin of the Museum of Comparative Zoology, 147:489-511.

Hook, R. W.
1983. *Colosteus scutellatus* (Newberry), a primitive temnospondyl amphibian from the Middle Pennsylvanian of Linton, Ohio. American Museum Novitates, no. 2270:1-41.
1989. Stratigraphic distribution of tetrapods in the Bowie and Wichita Groups, Permo-Carboniferous of north-central Texas. *In* R. W. Hook, ed., *Permo-Carboniferous Vertebrate Paleontology, Lithostratigraphy, and Depositional Environments of North-Central Texas.* Field Trip Guidebook No. 2, 49th Annual Meeting of the Society of Vertebrate Paleontology, Austin, Texas, 1989. Pp. 47-53.

Hook, R. W., and D. Baird
1986. The diamond coal mine of Linton, Ohio, and its Pennsylvanian-age vertebrates. Journal of Vertebrate Paleontology, 6:174-190.

Huene, F. von
1913. The skull elements of the Permian tetrapods in the American Museum of Natural History, New York. Bulletin of the American Museum of Natural History, 32:315-386.

Kutty, T. S.
1972. Permian reptilian fauna from India. Nature, 237:462-463.

Laerm, J. A.
1979. On the origin of rhipidistian vertebrae. Journal of Paleontology, 53:175-186.

Lane, H. H.
1945.   New Mid-Pennsylvanian reptiles from Kansas. Transactions of the Kansas Academy of Science, 47:381-390.
1946.   A survey of fossil vertebrates in Kansas: Part 3, the reptiles. Transactions of the Kansas Academy of Science, 49:289-332.

Langston, W.
1966.   *Limnosceloides brachycoles* (Reptilia: Captorhinomorpha), a new species from the Lower Permian of New Mexico. Journal of Paleontology, 40:690-695.

Lauder, G. V.
1980.   On the relationship of the myotome to the axial skeleton in vertebrate evolution. Paleobiology, 6:51-56.

Lewis, G. E., and P. P. Vaughn
1965.   Early Permian vertebrates from the Cutler Formation of the Placerville area, Colorado. Contributions to Paleontology, Geological Society Professional Paper 503-C (with a section on "Footprints from the Cutler Formation" by Donald Baird); 46 pages.

Maderson, P. F. A.
1975.   Embryonic tissue interactions as the basis for morphological change in evolution. American Zoologist, 14:141-172.
1983.   An evolutionary view of epithelial-mesenchymal interactions. *In* R. H. Sawyer and J. F. Fallon, eds., *Epithelial Mesenchymal Interactions*, pp. 215-242. New York: Praeger.

Maurer, F.
1892.   Der aufbau und die Entwicklung der ventralen Rumpfmuskulatur bei den urodelen Amphibien und deren Beziehung zu den gleichen Muskeln der Selachier und Teleostier. Morphologisches Jahrbuch, 18:76-179, plates 4-6.
1896.   Die Ventrale Rumpfmuskulatur einiger Reptilien. Festschrift zum Siebenzigsten Geburtstage von Carl Gegenbaur, Erster Band, S. 183-256.
1899.   Die Rumpfmusculatur der Wirbeltiere und die Phylogenese der Muskelfaser. Anatomie Hefte, Ergebnisse 9, T. 2, S. 692-819.

Mehl, M. G.
1912.   *Pantylus cordatus* Cope. Journal of Geology, 20:21-27.

Miner, R. W.
1925.   The pectoral limb of *Eryops* and other primitive tetrapods. Bulletin of the American Museum of Natural History, 51:145-312.

Moss, J. L.
1972. The morphology and phylogenetic relationships of the Lower Permian tetrapod *Tseajaia campi* Vaughn (Amphibia: Seymouriamorpha). University of California Publications in Geological Sciences, 98:1-72.

Nishi, S.
1916. Zur Vergleichenden Anatomie der eigentlichen (genuien) Rückenmusckeln. Morphologische Jahrbuch, 50, S:168-318.

Olson, E. C.
1936. The dorsal axial musculature of certain primitive Permian tetrapods. Journal of Morphology, 59:265-311.
1937. A mounted skeleton of *Labidosaurus* Cope. Journal of Geology, 45:95-100.
1947. The family Diadectidae and its bearing on the classification of reptiles. Fieldiana: Geology, 11:3-53.
1954. Fauna of the Vale and Choza: 9 Captorhinomorpha. Fieldiana, Geology, 10:211-218.
1962. Late Permian terrestrial vertebrates, U.S.A. and U.S.S.R. Transactions of the American Philosophical Society, 52:1-224.
1965. Relationships of *Seymouria, Diadectes*, and Chelonia. American Zoologist, 5:295-307.
1966. Relationships of *Diadectes*. Fieldiana, Geology, 14:199-227.
1970. New and little known genera and species from the Lower Permian of Oklahoma. Fieldiana: Geology, 18:359-434.
1976. The exploitation of land by early tetrapods. *In* A. d'A. Bellairs and C. B. Cox, eds., Linnean Society Symposium Series no. 3, *Morphology and Biology of Reptiles*, pp. 1-30.
1979. *Seymouria grandis* n. sp. (Batrachosauria: Amphibia) from the Middle Clear Fork (Permian) of Oklahoma and Texas. Journal of Paleontology, 53:720-728.
1980. The North American Seymouriidae. *In* L. L. Jacobs, ed., *Aspects of Vertebrate History*, pp. 137-152. Flagstaff: Museum of Northern Arizona Press.
1984. The taxonomic status and morphology of *Pleuristion brachycoelous* Case; referred to *Protocaptorhinus pricei* Clark and Carroll (Reptilia: Captorhinomorpha). Journal of Paleontology, 58:1282-1295.

Olson, E. C., and H. Barghusen
1962. Permian vertebrates from Oklahoma and Texas: Part I. Vertebrates from the Flowerpot Formation. Permian of Oklahoma (Olson and Barghusen). Part II. The osteology of *Captorhinikos chozaensis* (Olson). Oklahoma Geological Survey no. 59. 68 pages.

Olson, E. C. and J. R. Beerbower
1953. The San Angelo Formation, Permian of Texas, and its vertebrates. Journal of Geology, 61:389-423.

Olson, E. C., and J. G. Mead
1982. The Vale Formation (Lower Permian), its vertebrates and paleoecology. Texas Memorial Museum Bulletin 29; 46 pages.

Osawa, G.
1898. Beiträge zur anatomie der *Hatteria punctata*. Archiv. für Mikree Anatomie u. Ent., 51:481-691.

Oster, G., and P. Alberch
1982. Evolution and bifurcation of developmental programs. Evolution, 36:444-459.

Panchen, A. L.
1967. The homologies of the labyrinthodont centrum. Evolution, 21:24-33.
1977. The origin and evolution of tetrapod vertebrae. *In* S. M. Andrews, R. S. Miles and A. D. Walker, eds., Linnean Society Symposium Series no. 4, *Problems in Vertebrate Evolution*, pp. 289-318.
1980. The origin and relationships of the anthracosaur Amphibia from the Late Paleozoic. *In* A. L. Panchen, ed., Systematics Association Special Volume no. 15, *The Terrestrial Environment and the Origin of Land Vertebrates*, pp. 319-350. New York: Academic Press.

Panchen, A. L., and T. R. Smithson
1988. The relationships of the earliest tetrapods. *In* M. J. Benton, ed., Systematics Association Special Volume no. 35A, *The Phylogeny and Classification of the Tetrapods, Volume 1: Amphibians, Reptiles, Birds*, pp. 1-32. Oxford: Clarendon Press.

Parrington, F. R.
1967. The vertebrae of early tetrapods. Colloques Internationaux du Centre National de la Recherche Scientifique, Problèmes Actuels de Paléontologie (Evolution des Vertébrés), 163:269-279.

Peabody, F. E.
1952. *Petrolacosaurus kansensis* Lane, a Pennsylvanian reptile from Kansas. University of Kansas Paleontological Contributions, Vertebrata, article 1; 41 pages.

Penning, L.
1978. Normal movements of the cervical spine. American Journal of Roentgenology, 130:317-326.

Price, L. I.
1937. Two new cotylosaurs from the Permian of Texas. Proceedings of the New England Zoological Club, 41:97-102.

1940. Autotomy of the tail in Permian reptiles. Copeia, Number 2:119- 120.

Reisz, R. R.
1972. Pelycosaurian reptiles from the Middle Pennsylvanian of North America. Bulletin of the Museum of Comparative Zoology, 144:27- 61.
1975. Pennsylvanian pelycosaurs from Linton, Ohio and Nyrany, Czechoslovakia. Journal of Paleontology 49:522-527.
1977. *Petrolacosaurus*, the oldest known diapsid reptile. Science, 196:1091-1093.
1980. The Pelycosauria: a review of the phylogenetic relationships. *In* A. L. Panchen, ed., Systematics Association Special Volume no. 15: *The Terrestrial Environment and the Origin of Land Vertebrates*, pp. 553-592. New York: Academic Press.
1981. A diapsid reptile from the Pennsylvanian of Kansas. Special Publication of the Museum of Natural History, University of Kansas, no. 7; 74 pages.
1986. Pelycosauria. Handbuch der Paläoherpetologie, 17:1-102.

Reisz, R. R., D. S Berman, and D. Scott
1984. The anatomy and relationships of the Lower Permian reptile *Araeoscelis*. Journal of Vertebrate Paleontology, 4:57-67.

Ricqlès, A. de
1984. Remarques systématiques et méthodologiques pour servir à l'étude de la Famille des Captorhinidés (Reptilia, Cotylosauria, Captorhinomorpha). Annales de Paléontologie, 70:1-39.

Ricqlès, A. de, and P.Taquet
1982. La faune de Vertébrés du Permien supérieur du Niger, I. Le captorhinomorphe *Moradisaurus grandis* (Reptilia, Cotylosauria). Annales de Paléontologie, 68:33-106.

Romer, A. S.
1922. The locomotor apparatus of certain primitive and mammal-like reptiles. Bulletin of the American Museum of Natural History, 46:517-606.
1937. New genera and species of pelycosaurian reptiles. Proceedings of the New England Zoological Club, 16:83-96.
1944. The Permian cotylosaur *Diadectes tenuitectus*. American Journal of Science, 242:139-144.
1946. The primitive reptile *Limnoscelis* restudied. American Journal of Science, 244:149-188.
1947. Review of the Labyrinthodontia. Bulletin of the Museum of Comparative Zoology, 99:1-368.
1950. The nature and relationships of the Paleozoic microsaurs. American Journal of Science, 248:628-654.
1952. Late Pennsylvanian and early Permian vertebrates of the Pittsburgh-West Virginia region. Annals of the Carnegie Museum, 33:47-110.

1956. *Osteology of the Reptiles*. Chicago: University of Chicago Press. 772 pages.
1964. *Diadectes* an amphibian? Copeia, 4:718-719.
1966. *Vertebrate Paleontology*, 3rd ed. Chicago: University of Chicago Press. 468 pages.
1969. The cranial anatomy of the Permian amphibian Pantylus. Breviora, Number 314. 37 pages.

Romer, A. S., and L. I. Price
1940. Review of the Pelycosauria. Geological Society of America Special Papers, no. 28; 538 pages.

Sawin, P. B.
1946. Morphogenetic studies of the rabbit III. Skeletal variations resulting from the interactions of gene determining growth forces. Anatomical Record, 96:183-200.

Sawin, P. B., and I. B. Hull
1946. Morphogenetic studies of the rabbit II. Evidence of regionally specific hereditary factors influencing the extent of the lumbar region. Journal of Morphology, 78:1-26.

Seltin, R. J.
1959. A review of the family Captorhinidae. Fieldiana: Geology, 10:461-509.

Smithson, T. R.
1985. The morphology and relationships of the Carboniferous amphibian *Eoherpeton watsoni* Panchen. Zoological Journal of the Linnean Society, 85:317-410.

Stephens, T. D., and T. R. Strecker
1985. Radial condensation in the axis of the evolving limb. Evolution, 39:1159-1163.

Stevens, P. F.
1980. Evolutionary polarity of character states. Annual Review of Ecology and Systematics, 11:333-358.

Stovall, J. W.
1950. A cotylosaur from north central Texas. American Journal of Science, 248:46-54.

Stovall, J. W., L. I. Price, and A. S. Romer
1966. The postcranial skeleton of the giant Permian pelycosaur *Cotylorhynchus romeri*. Bulletin of the Museum of Comparative Zoology, 135:1-30.

Sumida, S. S.
1987. Two different forms in the axial column of *Labidosaurus* (Captorhinomorpha: Captorhinidae). Journal of Paleontology, 61:155-167.
1989a. The appendicular skeleton of the Early Permian genus *Labidosaurus* (Reptilia, Captorhinimorpha, Captorhinidae) and the hind limb musculature of captorhinid reptiles. Journal of Paleontology, 9:295-313.
1989b. Reinterpretation of vertebral structure in the Early Permian pelycosaur Varanosaurus acutirostris (Amniota, Synapsida). Journal of Vertebrate Paleontology, 9:419-426.

Taquet, M. P.
1969. Première découverte un Afrique d'un reptile captorhinomorphe (Cotylosaurien). C. R. Academe Scientifique Paris, serie D, 268:779-781.

Tartarinov, L. P.
1964. Order Placodontia: Subclass Lepidosauria. Osnovy Paleontologii, 12:332-338, 439-493.

Vaughn, P. P.
1955. The Permian reptile *Araeoscelis* restudied. Bulletin of the Museum of Comparative Zoology, 113:305-467.
1958. A specimen of the captorhinid reptile *Captorhinikos chozaensis* Olson, 1954, from the Hennessey Formation, Lower Permian of Oklahoma. Journal of Geology, 66:327-332.
1962. The Paleozoic microsaurs as close relatives of reptiles, again. American Midland Naturalist, 67:79-84.
1964. Vertebrates from the Organ Rock Shale of the Cutler Group, Permian of Monument Valley and vicinity, Utah and Arizona. Journal of Paleontology, 38:567-583.
1966. *Seymouria* from the Lower Permian of southeastern Utah, and possible sexual dimorphism in that genus. Journal of Paleontology, 40:603-612.
1969. Upper Pennsylvanian vertebrates from the Sangre de Cristo Formation of central Colorado. Los Angeles County Museum Contributions in Science, no. 164; 28 pages.
1970. Alternation of neural spine height in certain Permian tetrapods. Bulletin of the Southern California Academy of Sciences, 69:80-86.
1972. More vertebrates, including a new microsaur, from the Upper Pennsylvanian of central Colorado. Los Angeles County Museum Contributions in Science, no. 223; 30 pages.

Vjushkov, B. P., and P. K. Chudinov
1957. Discovery of a captorhinid in the Upper Permian of the U.S.S.R. Doklady Akademii Naus SSSR,112:523-526. (In Russian.)

Waddington, C. H.
1975.   *The Evolution of an Evolutionist*. Ithaca, N.Y.: Cornell University Press. 328 pages.

Wainwright, S. A., W. D. Biggs, J. D. Currey, and J. M. Gosline
1976.   *Mechanical Design in Organisms*. Princeton, N.J.: Princeton University Press. 423 pages.

Wake, D. B.
1970.   Aspects of vertebral evolution in the modern Amphibia. Forma et Functio, 3:33-60.

Wake, D. B., and M. H. Wake
1986.   On the development of vertebrae in gymnophione amphibians. Mémoires de la Société Zoologique de France, 43:67-70.

Watson, D. M. S.
1914.   Notes on *Varanosaurus acutirostris* Broili. Annals and Magazine of Natural History, ser. 8, 13:297-310.
1916.   On the structure of the brain-case in certain Lower Permian tetrapods. Bulletin of the American Museum of Natural History, 35:611-636.
1917.   The evolution of the tetrapod shoulder girdle and forelimb. Journal of Anatomy and Physiology, 52:1-63.
1918.   On *Seymouria*, the most primitive known reptile. Proceedings of the Zoological Society of London, 267-301.

Wellstead, C.
1985.   Taxonomic revision of the Permo-Carboniferous lepospondyl amphibian families Lysorophidae and Molgophidae. Ph.D. dissertation, McGill University, Montreal.

White, A. A., and M. M. Panjabi
1978.   *Clinical Biomechanics of the Spine*. Philadelphia: J. B. Lippincott Company. 534 pages.

White, T. E.
1939.   Osteology of *Seymouria baylorensis* Broili. Bulletin of the Museum of Comparative Zoology, 85:325-409.

Wiles, P.
1935.   Movements of the lumbar vertebrae during flexion and extension. Proceedings of the Royal Society of Medicine, 28:647-651.

Williston, S. W.
1909. New or little known Permian vertebrates, *Pariotichus*. Biological Bulletin, 17:241-255.
1910. New Permian reptiles; rhachitomous vertebrae. Journal of Geology, 18:585-600.
1911a. American Permian Vertebrates. Chicago: University of Chicago Press. 145 pages.
1911b. A new family of reptiles from the Permian of New Mexico. American Journal of Science, 31:378-398.
1912. Restoration of *Limnoscelis,* a cotylosaur reptile from New Mexico. American Journal of Science, 34:457-468.
1913. *Ostodolepis brevispinatus*, a new reptile from the Permian of Texas. Journal of Geology, 21:363-366.
1914. The osteology of some American Permian vertebrates. Contributions from Walker Museum, 1:107-162.
1916. (1) The osteology of some American Permian vertebrates II; (2) Synopsis of the American Permo-Carboniferous Tetrapods. Contributions from the Walker Museum, 1:165-236.
1917. *Labidosaurus* Cope, a Lower Permian cotylosaur reptile from Texas. Contributions of the Walker Museum, 2:45-57.
1925. *Osteology of the Reptiles*. Cambridge, Mass.: Harvard University Press. 300 pages.

Wyman, J.
1858. On some remains of batrachian reptiles discovered in the coal formation of Ohio. American Journal of Science, 25:158-164.

Plates

Plate 1. *Captorhinus aguti,* photomicrographs of thin sections. A, UCLA VP 3772; horizontal section near base of tall-type neural spine. Note the well-developed network of trabeculae. Anterior is to the left. B, UCLA VP 3770; sagittal section through neural arch and tall-type neural spine. The lower border is the roof of the neural canal. Anterior is to the right. C, UCLA VP 3764; horizontal section through low-type neural spine. Dark spot to the left is a mineral inclusion. Anterior is to the left. D, UCLA VP 3769; sagittal section through the neural arch and neural spine of a low-type neural spine. Note the lack of trabecular structures. Lower border is the roof of the neural canal. Anterior is to the left. E, UCLA VP 3762; horizontal section through the tip of a tall-type neural spine displaying well-developed trabeculae. Section is from a level somewhat more dorsal than that illustrated in A. The right posterolateral section of the spine has been broken away. Anterior is to the left. F, detail of region framed in lower right of E. Note the orientation of lacunae and vascular canals toward the posterolateral aspect of the neural spine. Anterior is to the left.

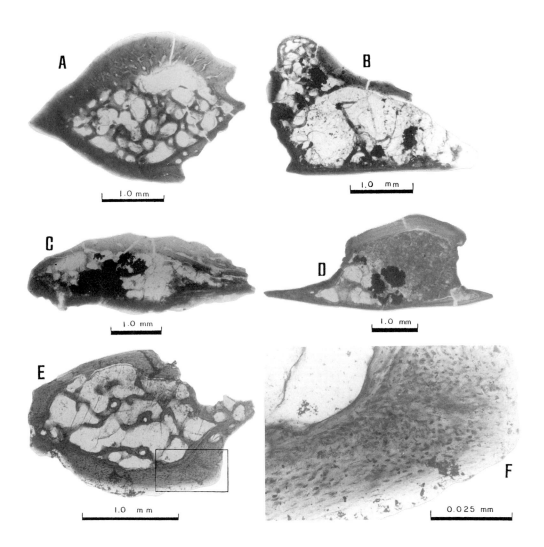

Plate 2. *Labidosaurus hamatus*. A-G, UCLA VP 3167. A, atlantal intercentrum, posterior aspect. B, atlantal intercentrum, ventral aspect, anterior facing top of page. C, left atlantal neural arch, medial aspect. D, left atlantal neural arch, lateral aspect. E, atlantal centrum, anterior aspect. Left rib articulation chipped and ventral keel slightly displaced. F, atlantal centrum, right lateral aspect. G, axial vertebra; neural spine, part of centrum, and axial rib. A portion of the hyoid apparatus and some matrix are attached to the specimen. H, FMNH UC726. Dorsal vertebrae in left lateral aspect of specimen that does not display alternation of neural spine height. I, FMNH UC726. Last three presacral and first sacral vertebra with sacral rib, left lateral aspect. J, FMNH UC726. Same specimen as (I), lower left, dorsal aspect; anterior to the left. Dorsal view shows differences in break patterns between tall- and low-spined vertebrae. K, MCZ 8923, dorsal vertebrae in left lateral aspect, showing alternation of neural spine height. L, UCLA VP 3491, juvenile; caudal vertebrae, left lateral aspect. Dark material is matrix. Lighter material surrounding matrix is plaster. Photographs reprinted with permission of the Journal of Paleontology.

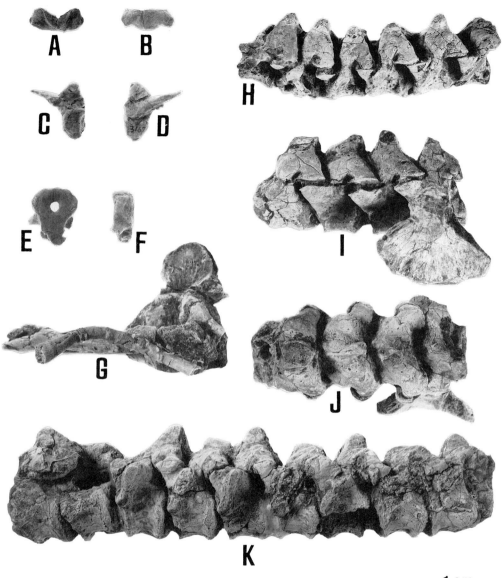

Plate 3. CM 47653, *Limnoscelis sp*. A, anterior dorsal vertebra (cf. 4th presacral), left lateral aspect. B, same as A, anterior aspect. C, mid-dorsal vertebra with low-type neural spine, anterior aspect. D, mid-dorsal vertebra with tall-type neural spine, anterior aspect. E, posterior dorsal vertebra with tall-type neural spine, anterior aspect. F, anterior caudal vertebra (cf. 3rd or 4th caudal), posterior aspect. Portion of right rib remains attached. G, caudal vertebra (cf. 7th or 8th caudal), right lateral aspect. H, right, mid-dorsal rib, dorsal aspect. I, left, anterior dorsal rib demonstrating spatulate distal expansion, dorsolateral aspect. J, right caudal rib (cf. 3rd or 4th caudal rib), dorsal aspect.

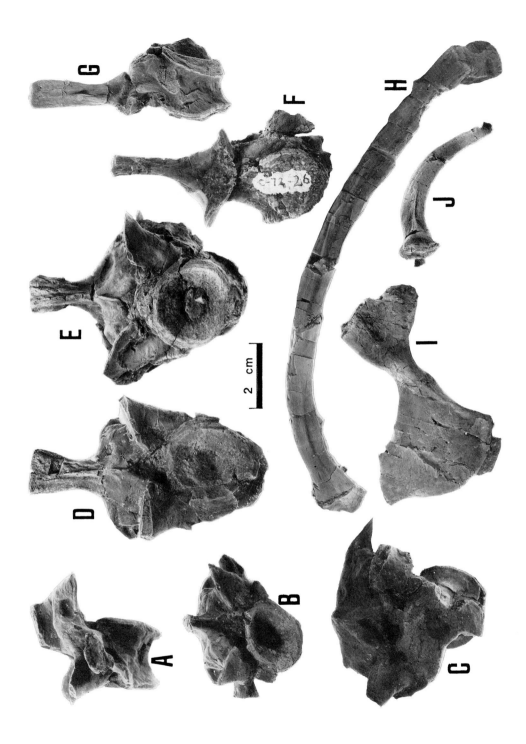

Other Volumes Available
University of California Publications in Zoology

Vol. 112. Ned K. Johnson. *Character Variation of Evolution of Sibling Species in the Empidonax Difficilis-Flavescens Complex (Aves: Tyrannidae).* ISBN 0-520-09799-5.

Vol. 114. Blair Csuti. *Type Specimens of Recent Mammals in the Museum of Vertebrate Zoology, University of California, Berkeley.* ISBN 0-520-09622-3.

Vol. 115. Peter B. Moyle et al. *Distribution and Ecology of Stream Fishes of the Sacramento-San Joaquin Drainage System, California.* ISBN 0-520-09650-9.

Vol. 116. Russell Greenberg. *The Winter Exploitation Systems of Bay-breasted and Chestnut-sided Warblers in Panama.* ISBN 0-520-09670-3.

Vol. 117. Marina Cords. *Mixed-Species Association of Cercopithecus Monkeys in the Kakamega Forest, Kenya.* ISBN 0-520-09717-3.

Vol. 118. Kevin de Queiroz. *Phylogenetic Systematics of Iguanine Lizards: A Comparative Osteological Study.* ISBN 0-520-09730-0.

Vol. 119. John E. Cadle. *Phylogenetic Relationships Among Advanced Snakes: A Molecular Perspective.* ISBN 0-520-09956-7.

Vol. 120. Kenneth A. Nagy and Charles C. Peterson. *Scaling of Water Flux Rate in Animals.* ISBN 0-520-09738-6.

Vol. 121. David A. Good. *Phylogenetic Relationships Among Gerrhonotine Lizards: An Analysis of External Morphology.* ISBN 0-520-09744-0.

Vol. 123. James L. Patton and Margaret F. Smith. *The Evolutionary Dynamics of the Pocket Gopher Thomomys bottae, with Emphasis on California Populations.* ISBN 0-09761-0.

University of California Press
Berkeley 94720

ISBN 0-520-09755